D0205910

Environmental Fate and Effects of Pesticides

ACS SYMPOSIUM SERIES **853**

Environmental Fate and Effects of Pesticides

Joel R. Coats, Editor
Iowa State University

Hiroki Yamamoto, Editor
Shimane University

**Sponsored by the
ACS Division of Agrochemicals**

American Chemical Society, Washington, DC

Library of Congress Cataloging-in-Publication Data

Environmental fate and effects of pesticides / Joel R. Coats, editor, Hiroki Yamamoto, editor ; sponsored by the ACS Division of Agrochemicals.

    p. cm.—(ACS symposium series ; 853)

    Includes bibliographical references and index.

    ISBN 0–8412–3722–0

    1. Pesticides—Environmental aspects. 2. Agricultural chemicals—Environmental aspects.

    I. Coats, Joel R. II. Yamamoto, Hiroki, 1947- III. American Chemical Society. Division of Agrochemicals. IV. Series.

TD196.P38E59   2003
628.5′29—dc21   2003048137

The paper used in this publication meets the minimum requirements of American National Standard for Information Sciences—Permanence of Paper for Printed Library Materials, ANSI Z39.48–1984.

PRINTED IN THE UNITED STATES OF AMERICA

Foreword

The ACS Symposium Series was first published in 1974 to provide a mechanism for publishing symposia quickly in book form. The purpose of the series is to publish timely, comprehensive books developed from ACS sponsored symposia based on current scientific research. Occasionally, books are developed from symposia sponsored by other organizations when the topic is of keen interest to the chemistry audience.

Before agreeing to publish a book, the proposed table of contents is reviewed for appropriate and comprehensive coverage and for interest to the audience. Some papers may be excluded to better focus the book; others may be added to provide comprehensiveness. When appropriate, overview or introductory chapters are added. Drafts of chapters are peer-reviewed prior to final acceptance or rejection, and manuscripts are prepared in camera-ready format.

As a rule, only original research papers and original review papers are included in the volumes. Verbatim reproductions of previously published papers are not accepted.

ACS Books Department

Contents

Indexes

Preface

The primary purpose of this book is to provide an update on research in the field of environmental impact of agricultural chemicals. As society becomes more concerned with non-target effects of pesticides and their residues in the environment, it is imperative that researchers develop a deeper and broader understanding of the types of effects, the quantities of the compounds that are biologically available, and the probability of serious impacts. Risk assessment can utilize information about the effects and the quantities to develop assessments of the likelihood for a given chemical, used in a specific place and time, to have deleterious effects on various species or ecological associations in the ecosystem.

The chemistry of agrochemicals is steadily growing more complex; new types of side effects are being detected; new analytical methodology is allowing for smaller quantities of the chemicals and their metabolites to be identified in the environment; regulatory agencies are developing more complicated guidelines and restrictions for testing and use of the new chemicals, as well as of the old ones. Overall, the landscape for discovery, development, and utilization of pest control chemicals has changed dramatically in the past half-century, and the need for reliable and meaningful environmental data and its interpretation has intensified greatly.

This book focuses on new approaches to the study of agrochemicals, including (1) newly developed experimental methods and novel analytical techniques; (2) important information on the environ-mental fate of recently developed active ingredients; and (3) advances made in refinement of risk assessment processes. This book includes some information on very specific compounds, but also presents broader views, such as an overview of the fate of pesticides in tropical ecosystems. The majority of the chapters are based on papers presented at the 2nd Pan-Specific Conference on Pesticide Science, although some authors were invited to write chapters after the conference. The ACS Division of Agrochemicals and the Pesticide Science Society of Japan were the primary organizers and sponsors of the conference. The authors are internationally preeminent authorities on the study of pesticides in the environment, and, by contributing chapters to this volume, the distribution of the information is wider and the impact will extend for a much longer time. The agrochemicals field is rapidly changing, and the research update presented herein is critically important for scientists in this and many related fields. We appreciate the time, effort, and expertise that the authors contributed to this book. The editors hope that it will be of value to numerous scientists, educators, consultants, and regulators. The book is dedicated to Rebecca J. Coats, with many thanks for her expert clerical assistance.

Joel R. Coats
Department of Entomology
Iowa State University
116 Insectary
Ames, IA 50011–0001
Email: jcoats@iastate.edu

Hiroki Yamamoto
Shimane University
1060 Nishikawatsu 690–0823
Japan
Email: yamahiro@life.shimaneu.ac.jp

Chapter 1

The Lysimeter Concept: A Comprehensive Approach to Study the Environmental Behaviour of Pesticides in Agroecosystems

F. Führ, P. Burauel, W. Mittelstaedt, T. Pütz, and U. Wanner

Institute of Chemistry and Dynamics of the Geosphere IV: Agrosphere, Forschungszentrum Jülich GmbH, D–52425 Jülich, Germany

The more proper and optimal use of pesticides will contribute considerably to provide the safe and reasonable priced food supply for the world population, which will increase during the next decade at a rate of about 80 millions per year predominantly in Asia, Africa and Latin America. In lysimeter experiments, the long-term behaviour of ^{14}C-labelled pesticides are investigated in reference agroecosystems under realistic climatic, cropping and soil conditions. Detailed information will be presented on the residue situation in soil and plants after repeated application of the ^{14}C-labelled herbicides methabenzthiazuron and fungicide anilazine. Special focus is given to uptake by treated and untreated rotational crops and on translocation and leaching including preferential flow phenomenon. The results are used as part of registration requirements and in addition can lead to practical recommendations for the improved use of pesticides in general.

1

World population will increase during the next decade at a rate of approximately 75 to 80 millions per year predominantly in Asia, Africa and Latin America. Especially in Asia the potential agricultural land is already in use by almost 90% and about 1% of valuable agricultural land is lost p.a. due to misuse and urban development. So we fully support the conclusion the Nobel Prize-winner Norman E. Borlaug (1) has drawn in his keynote lecture presented at the 15th world congress of Soil Science, Acapulco, Mexico in July 1994: *"Most of the increases in food production needed over the next several generations must be achieved through yield increases on land now under cultivation. Adaption of available agricultural technologies on the more-favoured lands will not only lead to economic development but will also do much to solve the serious environmental problems that come as a consequence of trying to cultivate lands unsuited to crop production. If the low income, food deficit nations are to feed themselves, the use of chemical fertilizers* – and we would like to add the more proper and optimal use of pesticides – *must be expanded in the developing countries over the next 20 years. "*

Pesticides have to be registered before they are allowed to be used and this means that they have to meet certain standards as laid down in the council directive concerning the placing of plant protection products on the market e.g. of the European Communities (2) where it is stated in article 4 that *"it is established, in the light of current scientific and technical knowledge that the compound is*

sufficiently effective;

has no unacceptable effect on plants or plant products;

has no unacceptable influence on the environment, having particular regard of the following considerations:

it's fate and distribution in the environment, particularly contamination of water including drinking water and ground water,

it's impact on non-target species.

The Council Directive of the European communities has defined certain criteria for the assessment of plant protection products in the registration procedure (2).The most prominent criteria concerning plant protection products involves their fate in the soil, the entry into the ground water, degradability and fate in the water/sediment system, volatility and behaviour in the air, and bioaccumulation and side effects on aquatic organisms, soil flora, earthworms, bees, birds, free-living mammals, and on beneficial organisms in general (3,4,5). Therefore, comprehensive information is needed on the environmental fate of pesticides.

For all of these criteria standardized laboratory or single species tests are developed and in use (3). But it is not an easy task to predict from such data the real situation under practical use of the compounds in agriculture. Therefore in the early seventies, in close cooperation with scientists from the Bayer AG, our institute adapted the lysimeter as a special tool to study the environmental behaviour of pesticides under realistic application conditions using ^{14}C-labelled compounds (6,7). Compared with laboratory techniques, the lysimeter experiment has the advantage that this design almost exactly reproduces the environmental conditions that occur in the corresponding field ecosystem. Growing crops in rotation as well as cultivation are in line with agricultural practice and the experiments can be maintained for many years. The results integrate the processes that are normally measured separately in the laboratory or with standardized experiments (7).

Some of the problems of field experimentation can be reduced in that soil monoliths from different sources can be grouped at the same site so it is possible to dispose different soil types to a particular climate. It is usually easier to install equipment to monitor environmental parameters at a lysimeter station than it is in the field, particularly when a large number of field sites are involved. In addition this approach includes all the processes affecting water movement in the agroecosystem and the special effects on pesticide uptake by the treated as well as the untreated rotational plants, on volatilization, on pesticide transport in soil and on the residue situation in the soil in general (7,8).

Undisturbed soil monoliths with a profile depth of 1.10 m were removed from the selected field site with the aid of stainless steel cylinders and inserted in stainless steel containers firmly embedded in the soil. More details are given by Steffens et al. (6). These lysimeters with a cultivated area of 0.5 or 1.0 m^2 are surrounded by control areas planted with the same crop. The experiments with ^{14}C-labelled pesticides were conducted in accordance with good agricultural practice. Fertilization and complementary plant protection measures were closely coordinated with agricultural practice. Natural precipitation as well as soil/air temperature and air humidity and soil moisture in different soil layers were recorded continuously.

The soil used for the experiments was removed from a 7.5 ha field. This soil is an orthic luvisol (Parabraunerde), a clayey silt derived from loess and widespread in the Federal Republic of Germany, and represents a fertile soil mainly used for agriculture. A second arable soil, a sandy soil (gleyic cambisol), was also available in order to include a more water-permeable soil type in the experimental programme, particularly for questions of the leaching behaviour of pesticides in accordance with the lysimeter guideline of the German authorities (9,10).

The following data, presented in four poster contributions to this conference, will demonstrate certain advantages of the lysimeter concept using ^{14}C-labelled pesticides to study their long-term environmental behaviour in agroecosystems, and in particular their binding and bioavailability in soil and carryover to untreated rotational crops.

4

Long-Term Lysimeter Studies With Repeated Application of Anilazine and Methabenzthiazuron

Anilazine (2,4-Dichloro-6-(2-chloroanilino)-1,3,5-triazine) is the active ingredient of the fungicide Dyrene®. Incubation studies with ^{14}C-labelled anilazine revealed that it tends to build up high quantities of non extractable residues in soil (11,12). Therefore a long term lysimeter study was carried out to investigate the binding processes and in particular the bioavailability in soil under outdoor conditions. [Phenyl-U-^{14}C]anilazine (Fig. 1) was applied as a Dyrene®-WP formulation in June 1985, followed by 4 repeated applications in 1986 through 1989 (13). Anailazine was applied directly onto the soil in-between the crop rows instead of foliar application as it is in agricultural practice simulating the maximum load reaching the soil via different fungicide applications. Two 1-m^2 lysimeters with 1.10 m undisturbed soil profile (6) were sprayed at a rate of 2 and 4 kg anilazine ha^{-1} and year, respectively (Table I). The soil was an orthic luvisol (Table II) and cultivated with cereal crops (Table I) during the five years of application. In 1985 through 1988 soil sampling was carried out by removing 0.15 m^2 of soil always 100 days after aplication. Soil

Test substance 1 (fungicide)
[**phenyl-U-^{14}C]anilazine** (2,4-Dichloro-6-(2-chloroanilino-U-^{14}C)-1,3,5-triazine)

Solubility: 0.008 g L^{-1} in water (20°C)

Anilazine:	R_1, R_2 = Cl
Dihydroxy anilazine:	R_1, R_2 = OH
Mono hydroxy anilazine:	R_1 = OH, R_2 = Cl

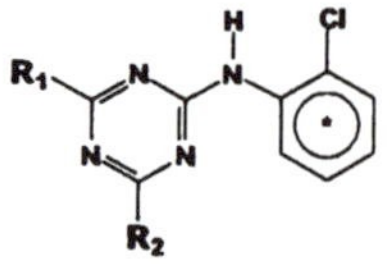

Test substance 2 (herbicide)
[**phenyl-U-^{14}C]methabenzthiazuron (MBT)** (1-Benzothiazol-2-yl-1,3-dimethyl urea)

Solubility: 0.059 g L^{-1} in water (20°C)

Methabenzthiazuron:	R_1 = CH$_3$
Demethyl methabenzthiazuron:	R_1 = H
* = Labelling Position	

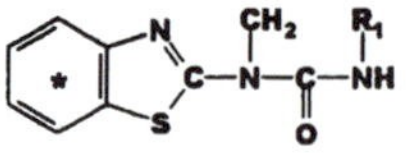

Figure 1. ^{14}C-labelling test substances and major metabolites.

® Registered trademark, Bayer AG Leverkusen, Germany

Table I. Crops/Rotation and Active Ingredient Application (kg ha⁻¹)

	[^{14}C]Anilazine		[^{14}C]Methabenzthiazuron	
Year	Crop	kg ha⁻¹	Crop	kg ha⁻¹
1985	winter wheat	2&4	-	
1986	winter barley	2&4	-	
1987	winter rye	2&4	-	
1988	oat	2&4	-	
1989	winter wheat	2&4	winter wheat	2.5
1990	winter barley	-	winter barley / oat	-
1991	winter rye	-	sugar beet	-
1992	sugar beet	-	oat	-
1993	winter wheat	-	winter wheat	3.2
1994	oat	-	sugar beet	-
1995	bare soil	-	winter wheat	3.1
1996	winter wheat	-	corn	-
1997	winter barley	-	clover, not harvested	-
1998	bare soil	-	corn	-
1999	winter wheat	-	winter wheat	-

Table II. Physico-Chemical Data of the Loess Soil

Soil:	Loess (clayey silt, orthic luvisol)				
	pH (KCl)	C_{org} [%]	Sand [%]	Silt [%]	Clay [%]
A_p	6.9	1.2	6.4	78.2	15.4
Al	6.9	0.4	1.0	77.1	21.9

6

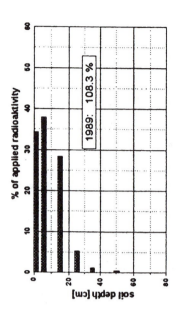

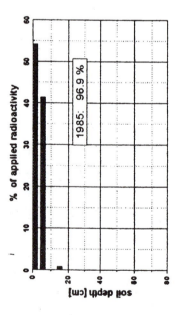

0-10 cm soil layer

1996	1998	1999
22.3 %	20.7 %	19.9 %

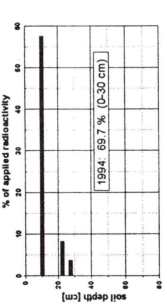

Figure 2. Residual radiocarbon in soil after 5 repeated applications of [phenyl-U-^{14}C]amilazine: 2 kg ha^{-1} in 1985 to 1989. (Total radioactivity applied = 100%).

samples were subdivided into 0-2.5 cm, 2.5-10 cm, 10-20 cm and 20-30 cm fractions for further analysis. Since 1989 soil cores were taken.

Due to the strong binding almost the total applied radioactivity could be recovered each year which, consequently means that no residue plateau (steady state between decomposition and application rate) was built up during five successive years of application (Fig. 2). Less than 3 % of the applied radioacitvity was translocated below 30 cm. With moderate extraction procedures (acetone/0.1 M $CaCl_2$ and dichloromethane) a total of 13 to 21 % of the applied ^{14}C was extractable in the 0-30 cm soil layer. The main metabolites in the organic and aqueous phases were monohydroxy- and dihydroxy-anilazine and only traces of anilazine could be characterized in the upper soil layers by radio thin-layer chromatography. The major part of the residual ^{14}C was associated with the humin fraction, followed by the fulvic acids fraction which gained more importance in the 20-30 cm soil layer. Therefore efforts were undertaken to elucidate binding mechanisms of anilazine to the humic substances as presented in the contribution of Wais et al. (14,15) and in this paper.

The treated plants contained at harvest, with the exceptions of the first treated winter wheat (1985) and oat (1988), less than 0.5 % of the applied ^{14}C-activity with the far highest contents always found in the straw (Fig. 3). This portion was reduced to less than 0.1% detected in the succeeding untreated rotational crops demonstrating the drastic reduced bioavailability of the soil-bound residues. It can be assumed from detailed studies with other ^{14}C-labelled pesticides (16) that a certain portion of this radiocarbon was assimilated via the leaves in the form of the mineralization end product of the phenyl ring carbon, the $^{14}CO_2$.

After the first anilazine application only traces (0.005 %) of ^{14}C-activity applied were leached out of the 110-cm soil profile with about 18 % of the annual precipitation (Fig. 4). During the second and third vegetation period the leachate increased to about 28 % of the annual precipitation and 0.09 % (1986/87) and 0.18 % (1987/88) of the applied ^{14}C-activity were detected in the leachate. During the succeeding vegetation periods (1990 through 1999) the annual leachate amounted up to 40 % of the precipitation and a maximum of 0.3 to 0.4 % of the applied ^{14}C-activity was detected in the leachate in 1992 with a decreasing trend starting in 1994. The total discharge of radioactivity until 1999 amounted to 1.9 % of the applied radioactivity. The $^{14}CO_2$-content rose steadily to 20 % of the radioactivity in the leachate, indicating ring cleavage of the active ingredient anilazine. Neither anilazine nor the major metabolites monohydroxy- and dihydroxy-anilazine were detected in the leachate.

In 1994 the radiocarbon remaining in the soil profile (2 kg anilazine ha^{-1}) represented 70 % of the total applied ^{14}C during the years 1985 till 1989 (Fig.

9

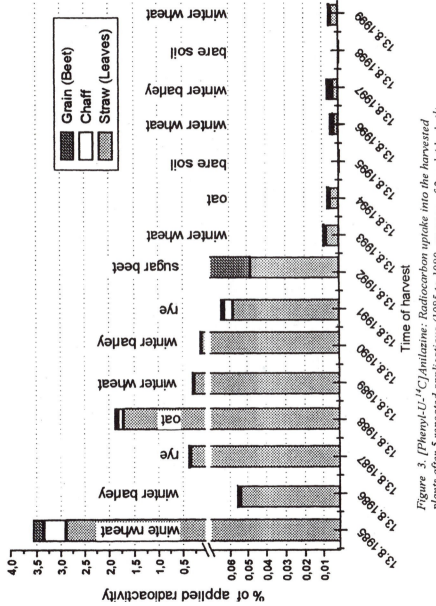

Figure 3. [Phenyl-U-¹⁴C]Anilazine: Radiocarbon uptake into the harvested plants after 5 repeated applications (1985 to 1989, average of 2 and 4 kg ha⁻¹).

10

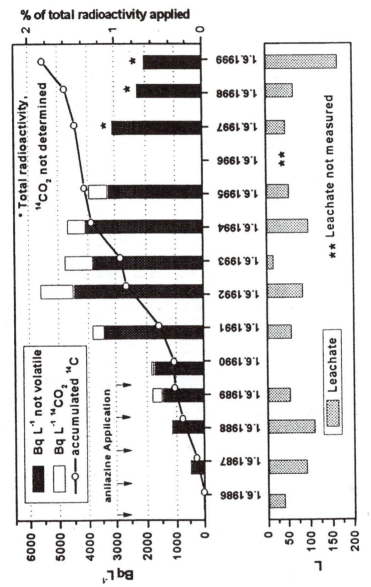

Figure 4. Anilazine: Annual ¹⁴C- and ¹⁴CO₂-concentrations and the accumulated ¹⁴C-activity in the leachate, calculated for the remaining area (Ø of 2 & 4 kg ha⁻¹).

2). In 1996 radioactivity in the upper 10 cm amounted to 22.3 % applied, and this portion was further reduced to just 19.9% in 1999, demonstrating that the fixed or bound phenyl ring of the anilazine molecule is finally drawn into the turnover of soil organic matter constituents at a higher rate than the stable soil organic matter itself.

Methabenzthiazuron: Over a period of 10 years 3 long-term experiments with the herbicide [phenyl-U-^{14}C]methabenzthiazuron (Fig. 1), active ingredient in Tribunil® applied post-emergence or pre-emergence to winter wheat, have been conducted in lysimeters filled with an undisturbed soil core of an orthic luvisol (6). With these studies (17-21) the most comprehensive information on the advantages of lysimeter studies in elucidating the environmental behaviour of pesticides have been collected. One lysimeter received spray applications (WP 67) 3 times pre-emergence to winter wheat in fall 1988, 1992 and 1994 simulating the practical Tribunil® application within typical rotations.

As demonstrated in Fig. 5 the treated plants took up about 1 % of the applied ^{14}C. The rotational crops immediately following the treated winter wheat were exposed to relatively high concentration of methabenzthiazuron residues during the first weeks of growing due to the intensified turnover of methabenzthiazuron in the soil solution (20; 21). However, only very small amounts are taken up as compared to the uptake of MBT residues by the treated crops with the only exception of corn (Fig. 5). This demonstrates that in lysimeter experiments the rotational untreated crop is integrating the uptake situation over time in dependence of the physico-chemical behaviour of the respective pesticide reflecting the turnover and biological availability in dependence of soil and climatic factors, too. This detailed information is an essential pre-requisite for assessing the exposure of e.g. soil organisms in order to link an observed effect to the bioavailability of a potential active ingredient (22).

After the first pre-emergence application of the methabenzthiazuron almost all the radioactivity remaining in the soil could be recovered in the upper 5 cm of the plough layer (Fig. 6). After simulated ploughing a more even distribution in the plough layer can be observed after the repeated application. Due to degradation and especially mineralization (23) on average about 85 % of the applied radioactivity remained in the soil at harvest time about 9 month after spray application. As in agricultural practice immediately after harvest of the treated winter wheat, the introduction of plant residues in the form of straw into the upper 0-10 cm of the soil stimulated microbial activities and hence lead to a dramatic increase in methabenzthiazuron degradation to the major metabolite demethyl methabenzthiazuron and to the formation of bound residues (21,23).

® Registered trademark, Bayer AG Leverkusen, Germany

12

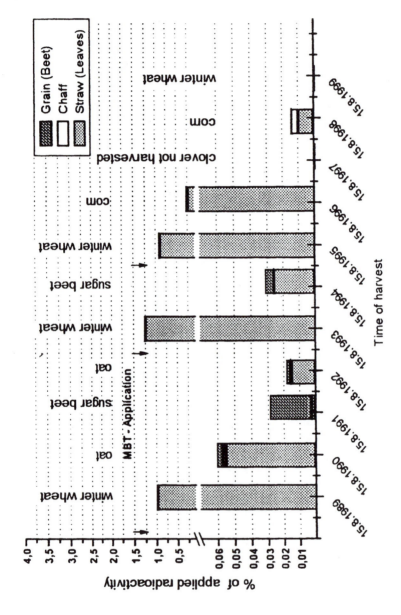

Figure 5. Methabenzthiazuron: Radiocarbon uptake into in the harvested plants after 3 pre-emergence applications in fall 1988, 1992 and 1994

Depending on the crops the amounts of leachate vary considerably. As shown in Fig. 7 the total ^{14}C occurring in the leachates over the period of 10 years is reaching a total of 0.23 % of the total radioactivity applied with the 3 sprayings. The ^{14}C recovered as carbonates in the leachate (up to 46 % of the radiocarbon in the leachate) demonstrate the ring clevage of the active ingredient methabenzthiazuron.

These results demonstrate that lysimeter experiments with ^{14}C-labelled pesticides, repeatedly applied in accordance with good agricultural practice, yield valuable information on the long-term behaviour of pesticides in agroecosystems. Lysimeter studies allow the design of detailed experiments to elucidate certain questions in connection with residue formation in soil, carryover into untreated succeeding crops, leachate transfer and translocation into deeper soil layers in particular (7).

Fate, Mobility and Bioavailability of Benazolin-Ethyl

A lysimeter study with the herbicide benazolin-ethyl applied to two soil types was undertaken (Table III). In order to monitor water flow in the lysimeters potassium bromide as an inert water tracer was applied simultaneously with the herbicide. Benazolin-ethyl is a broad spectrum herbicide mainly used for weed control in cereals and oil seed rape. Technical grade [benzene-U-^{14}C-]benazolin-ethyl and metabolites (as reference substances) were supplied by AgrEvo Company. The chemical names of the active ingredient and its major metabolites are given in Figure 8.

^{14}C-benazolin-ethyl was applied to four lysimeters (1 m^2 surface area, undisturbed soil monoliths of 1.1 m depth) at rates of 0.68 kg ha^{-1} for the KRA-soil and 0.75 kg ha^{-1} for the MER-soil (Table III) and bromide at a rate of 670 kg Br$^-$ ha^{-1}. No crop was planted in the first year. After 15 months corn was planted (14 plants per m^2, variety *Fanion*). Soil samples from one lysimeter of each soil type were collected monthly in the first six months and 13 months after the application. At each time point, six soil cores of 3.5 cm diameter and 30 cm depth were taken and divided into seven soil layers. Total residual radioactivity per soil layer was calculated on the basis of average soil densities determined during sampling. Pooled soil samples were extracted with simulated soil solution (0.01 M CaCl$_2$) for 24 h, followed by an extraction with acetone for 1 h and ethylacetate for 1 h (rotary shaker at 120 rpm, soil/solution ratio 1:2). Radioactivity in the extracts was measured by liquid scintillation counting and extracts were further analysed by radio-thin-layer chromatography with co-chromatography of non-labelled reference compounds. Residual radioactivity in soil was combusted for $^{14}CO_2$ determination. The other two lysimeters, one of each soil type, were equipped with devices to measure soil moisture content and temperature in various soil depths. These lysimeters were not disturbed by soil sampling. The leaching water was analysed for bromide, parent compound and metabolites.

After 3, 6 and 13 months of experimental time the amount of residual radioactivity was reduced from 90 to 43 and 35 % for KRA-soil and from 83 to

14

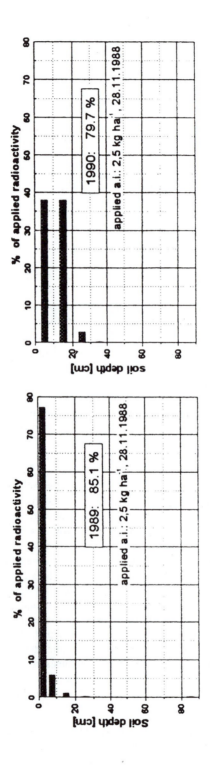

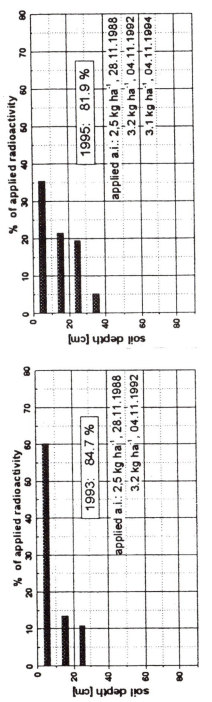

Figure 6. Residual radiocarbon in soil after 3 pre-emergence applications of MBT in 1988, 1992 and 1994. (Total radioactivity applied = 100 %

16

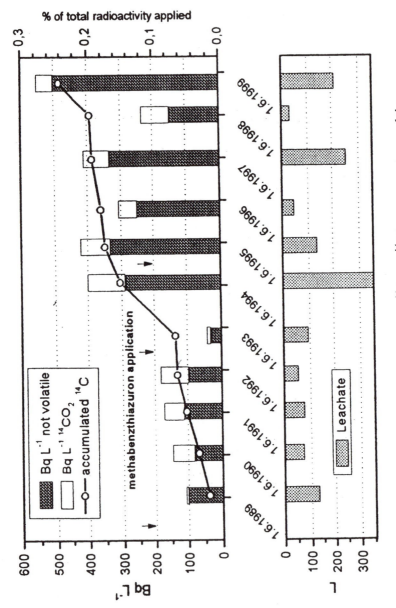

Figure 7. *Methabenzthiazuron: Annual* ^{14}C- *and* $^{14}CO_2$-*concentrations and the accumulated* ^{14}C-*activity in the leachate (Applied* ^{14}C-*activity = 100%).*

Parent: **Benazolin-ethyl** (ethyl 4-chloro-2-oxo-benzothiazolin-3-yl acetate)
Emperical formula: $C_{11}H_{10}ClNO_3S$
Molecular weight: 271.7
Melting point: 79°C
Solubility: 0.047 g/L in water
Structural formula:

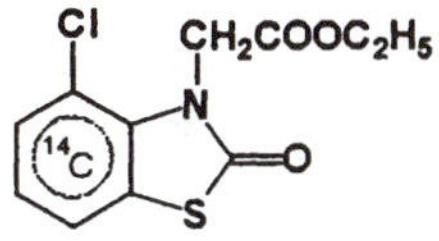

Metabolite: **Benazolin** (4-chloro-2-oxobenzothiazolin-3-yl acetic acid)
Emperical formula: C9H6ClNO3S
Molecular weight: 243.7
Melting point: 193°C
Solubility: 0.6 g/L in water
Structural formula:

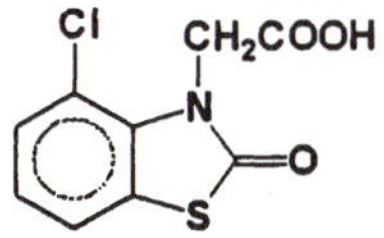

Metabolite: **BTS 18753** (4-chlorobenzothiazolin-2-one)
Emperical formula: C7H4ClNOS
Molecular weight: 185.6
Structural formula:

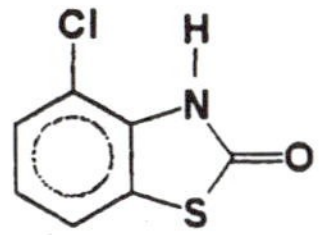

Figure 8. ^{14}C-labelled test substance and major metabolites

Table III Characteristics of the Gleyic Planosol from Krauthausen (KRA) and the Orthic Luvisol from Merzenhausen (MER), North Rhine-Westphalia in the Top Soil (A_p horizon).

	Horizon	pH (CaCl₂)	C_{org} [%]	Sand [%]	Silt [%]	Clay [%]
KRA-soil	A_p	7.2	1.3	8.8	73.3	17.9
MER-soil	A_p	7.2	1.2	6.4	78.2	15.4

43 and 44 % for a MER-soil respectively. Only traces of radioactivity was detected below 15 cm of depth at all samplings. Generally more than half of the residual radioactivity was not extractable. These findings are supported by Leake (24). Three months after the application of benazolin-ethyl more than 95 % of the extractable radioactivity could be characterized as benazolin (45 to 50 % of applied radioactivity) and 3 to 6 % as BTS 18753. After six months detectable quantities of benazolin dropped to between 4 and 9 % and BTS 18753 to an amount of 5 % of applied radioactivity for both soil types.

After 13 months benazolin and the metabolite BTS 18753 represented 2.1 % and 6.2 % applied radoactiviy in the MER-soil and 0.02 % and 2.7 % in the KRA-soil, respectively. At the same sampling point about 90 % of detectable residual radioactivity was still located in the top 20 cm of both soil monoliths. Benazolin-ethyl could neither be detected in the soil samples nor in the drainage water. This can be explained by its rapid hydrolysis to benazolin. Leake et al. (25) reported a half life for benazolin-ethyl in soil of < 1 day.

Figure 9 shows the bioavailability of ^{14}C-residual radioactivity to corn planted 15 months after the application of benazolin-ethyl. The total production of plant biomass varied between 1.6 to 1.7 kg dry matter per m². The uptake of residual radioactivity into the dry matter of the plants was correlated to the fraction of desorbable residual radioactivity in the top soil layer as a measure of its potential bioavailability. Desorbable radioactivity in the MER-soil was 2 to 4 times higher compared to the KRA-soil.

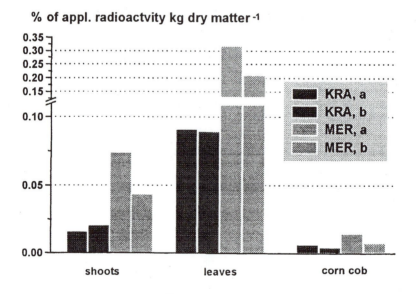

Figure 9. Bioavailability of ^{14}C-residual radioactivity to corn planted 15 months after application of benazolin. Single values for the parallel lysimeters a and b.

In Figures 10 and 11 the breakthrough curves of the ^{14}C radioactivity as well as the inert water tracer bromide are shown. During the experimental time of 17 months a sum of 1 to 3 % of applied radioactivity could be detected in the leachates (Fig. 10). Benazolin could be characterized only in the first 100 to 150 L of drainage water (~ LP1, Figure 10) which can be explained by preferential flow phenomenon supported by the simultaneous detection of bromide (Fig. 11) as well as benazoline (Fig. 10) in the leaching water shortly after application. The occurence of benazolin in the drainage water during the first weeks after application was found coincidently with high amounts of desorbable benazolin residues still present in the plough layer of both soil monoliths with 2/3 of analysed benazolin in the desorbable fraction.

The Formation of Non-Extractable Anilazine Residues in Soil

As seen in the lysimeter experiment the dissipation of the fungicidal component [phenyl-U-^{14}C]anilazine is mainly characterized by the formation of high amounts of non-extractable residues (Fig. 12). Most of this non-extracted radiocarbon is located in the fraction of the humins.

In order to characterize the chemical nature of these bound residues, ^{13}C-NMR spectroscopy can be used (14,15). The ^{13}C-enriched compound [triazine-U-^{13}C]anilazine was incubated for 20 days in an artificial, ^{13}C-depleted soil. This artificial soil was derived from a humification study of ^{13}C-depleted corn straw in an incinerated soil exposed to constant temperature and soil moisture regime over a period of 26 weeks (15). After the incubation with ^{13}C-anilazine the soil was extracted exhaustively with organic solvents. The humic substances of the extracted soil were fractionated by extraction with 0.5 M NaOH with subsequent pH lowering. The isolated humic acids were investigated using liquid ^{13}C-NMR spectroscopy (inversed gated decoupling). The spectra (Fig. 13) were compared with NMR spectra of the dihydroxy- and the dimethoxy-derivative of anilazine (2,4-Dihydroy-6-(2-chloroanilino)-1,3,5-triazine, respectively, 2,4-Dimethoxy-6-(2-chloroanilino)-1,3,5-triazine). The dimethoxy-derivative has characteristic chemical shifts at 166.4-166.9 ppm and at 172.0-172.8 ppm. The same chemical shifts can be observed in the spectra of anilazine bound to humic acids (Fig. 13). This proves covalent bonding of non-extracted residues (-C-O-C-binding) to the humic acids fraction.

There are several possible dissipation pathways which can occur after application of a pesticide. The soil, being sooner or later the ultimate sink of each compound, plays a central role in the overall environmental behaviour of pesticides. Due to processes in soil an applied pesticide might become fixed to the matrix as demonstrated with the results from the anilazine experiment. The quality of this fixation ranges from weak physi-sorption to strong interactions which turn the compounds non-extractable. According to the definition of the IUPAC (26) and the "authorization" directive 91/414/EEC (2) *"non-extractable residues (occasionally referred to as "bound residues") in soil are chemical*

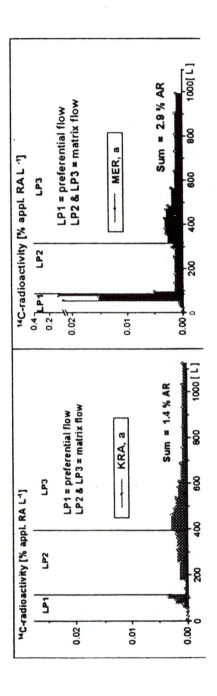

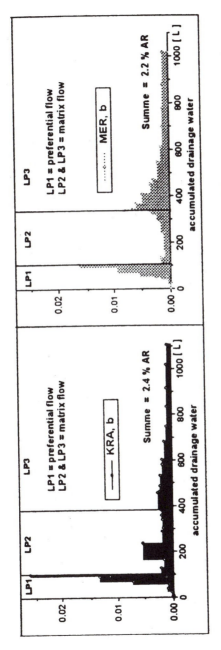

Figure 10. Breakthrough curves of ^{14}C in KRA (left, a, b) and MER (right, a, b) lysimeters, 17 months of monitored experimental time. LP = leaching period. Applied radioactivity = 100 %

22

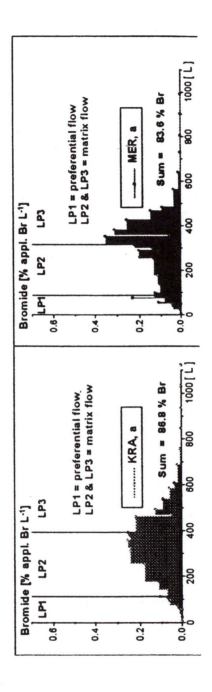

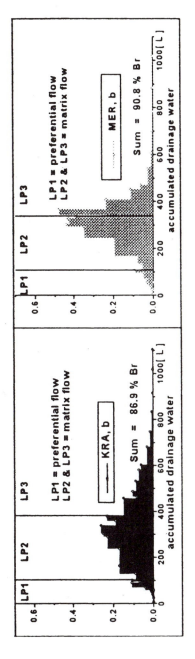

Figure 11. Breakthrough curves of the inert water tracer Br⁻ in KRA (left, a, b) and MER (right, a, b) lysimeters, 17 months of monitored experimental time LP = leaching period (Applied bromide = 100 %)

24

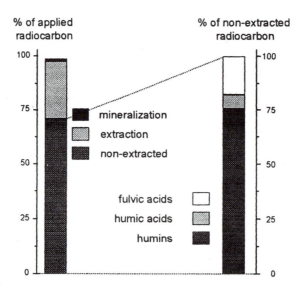

Figure 12. Degradation properties of anilazine and the distribution of non-extracted residues after 28 days of incubation in an orthic luvisol.

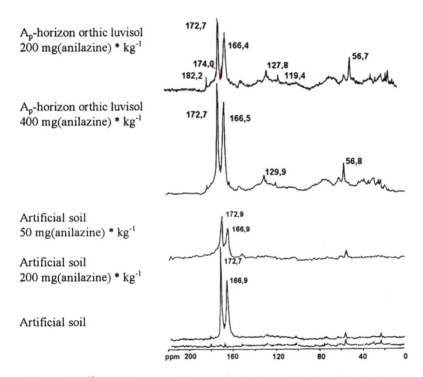

Figure 13. ^{13}*C-NMR spectra of humic acids of an orthic luvisol and of an artificial soil which were incubated with 50-400 mg*kg^{-1} [triazine-U-^{13}C]anilazine: δ = 166.4-166.9 ppm C-atom of the triazin-chloranilin bond; δ = 172.0-172.8 ppm C-atoms of the methoxy-substituted triazin (WAIS, 1997)*

species (active ingredient, metabolites and fragments) originating from pesticides, used according to good agricultural practice, that are unextracted by methods which do not significantly change the chemical nature of these residues, but which remain in soil. These non-extractable residues are considered to exclude fragments recycled through metabolic pathways leading to naturally occuring products. Calderbank (27) changed this definition by emphasizing investigations of the bioavailability of bound residues.

In the keynote of the workshop on bound residues in soil, organised by the Deutsche Forschungsgesellschaft (DFG), Führ ET AL. (28) proposed a modified definition of these non-extractable residues: *"Bound residues represent compounds in soil, plant or animal which persist in the matrix in the form of parent substance or its metabolite(s) after extractions. The extraction method must not substantially change the compounds themselves or the structure of the matrix. The nature of the bond can be clarified by matrix altering extraction methods and sophisticated analytical techniques. In general the formation of bound residues reduces the bioaccessibility and the bioavailability significantly."*

Conclusion

On the basis of the results of extensive long-term lysimeter experiments with the 3 ^{14}C-labelled pesticides demonstrated here and supported by data the Institute of Radioagronomy published in over 200 papers covering about 50 compounds combining microecosystem, pot, lysimeter and field experiments the following conclusion can be drawn: The soil constitues the main sink for all pesticides applied. Generally pesticides and their metabolites lose their bioactivity, and in many cases also their identity due to processes of degradation, sorption, fixation, sequestration and binding. This leads to varying residence times of the carbon from the molecular structures of pesticides, especially as a function of the soil carbon structures and the bond interactions goverened by additional soil and climatic factors. During the continuous turnover of native soil carbon non-extractable, bound residues may be released in small amounts and undergo further transformation, but mere ageing of the residues in soil leads already to a considerable reduction in their biological availability to plant roots. Lysimeter experiments yield valuable information on the bioavailability of these residues to treated crops and untreated rotational crops. The residue situation in soils and treated as well as untreated rotational plants should be used to illustrate the analytical detectable residue situation versus the bioavailability to plants. It is well demonstrated that depending on the application situation (e.g. seed dressing, pre- or post-emergence spraying) a different compartmentalization of residues occurs in soils. This detailed information is an essential prerequisite for assessing the exposure of soil organisms in order to link an observed effect to the occurrence of a potential active ingredient. The results of such long-term lysimeter studies as demonstrated in this contribution are used as part of registration requirements (9;10). But above all, they also contribute to improve and optimize the use of pesticides in general. As demonstrated, lysimeter experiments with ^{14}C-labelled pesticides yield detailed information on the occurance of residues in the soil solution and on the leaching and translocation behaviour of active ingredients and metabolites.

Acknowledgements:

Part of this research was supported by Agro-Evo, Germany and EU contract EV5V-CT92.-0214, and in particular by the Agricultural Centre Monheim, Bayer AG, Germany

References

1. Borlaug, N.E.: *Feeding a Human Populatin that Increasingly Crowds a Fragile Planet.* Keynote Lecture, 15. World Congress of Soil Science, Acapulco, Mexico, July 1994.
2. Anon. *Council Directive Concerning the Placing of Plant Protection Products on the Market (91/414/EEC).* Official Journal of the European Communities, 1991; No.L230/1
3. Anon. *Criteria for Assessment of Plant Protection Products in the Registration Procedure.* Mitt. Biologische Bundesanstalt für Land- und Forstwirtschaft 285, Berlin-Dahlem, Kommissionsverlag Paul Parey, Berlin and Hamburg, 1993; 1-125
4. G.I.F.A.P. *Environmental Criteria for the Registration of Pesticides.* Technical Monograph No. 3. Brussels, Belgium, 1990
5. Anon., *Data Requirements and Criteria for Decision-Making in the European Union and the Federal Republic of Germany for the Authorization Procedure of Plant Protection Products.* Mitt. Biologische Bundesanstalt für Land- und Forstwirtschaft 358, Berlin-Dahlem, Kommissionsverlag Paul Parey, Berlin and Hamburg 1998; 1-158
6. Steffens, W.; Mittelstaedt, W.; Stork, A.; Führ, F. In: *Lysimeter Studies of the Fate of Pesticides in the Soil.* Führ, F., Hance, R.J., Eds.; BCPC

28

Monograph No 53, The Lavenham Press Ltd., Lavenham, Suffolk, **1992**; 21-34

7. Führ, F.; Burauel, P.; Dust, M.; Mittelstaedt, W.; Pütz, T.; Reinken, G.; Stork, A. In: *The Lysimeter Concept: The Environmental Behaviour of Pesticides*, Führ, F. Hance, R.J. Plimmer, J.R. and Nelson, J.O. (EdsACS Symposium Series 699, Washington **1998**; 1-20

8. Hance, R.J.; Führ, F. In: *Lysimeter Studies of the Fate of Pesticides in the Soil.* Führ, F.; Hance, R.J., Eds., BCPC Monograph No. 53, The Lavenham Press Ltd., Lavenham Suffolk, **1992**; 9-18

9 Anonymous: *Lysimeteruntersuchungen zur Verlagerung von Pflanzenschutzmitteln in den Untergrund.* Biologische Bundesanstalt für Land- und Forstwirtschaft: Richtlinien für die Prüfung von Pflanzenschutzmitteln im Zulassungsverfahren, Saphir-Verlag, Braunschweig, **1990**; Part IV, 4-3, 1-11

10. Nolting, H.G.; Schinkel, K. In: *The Lysimeter Concept: Environmental Behaviour of Pesticides;* Führ, F.; Plimmer, J.R.; Hance, R.J.; Nelson, J.O.; Eds., ACS Symp. Ser. 699, Oxford University Press, New York; **1998**; pp 238-245

11. Mittelstaedt, W.; Führ, F.; Kloskowski, R. *J. Environ. Sci.Health* B, **1987**; 22, 491-507

12. Heitmann-Weber, B.; Mittelstaedt, W.; Führ, F. *J. Environ.Sci. Health* B, **1994**; 29, 247-264

13. Mittelstaedt, W.; Führ, F. In: *Pesticide Bound Residues in Soil.* Report 2 of the Senate Commission for the Assessment of Chemicals Used in Agriculture. Wiley-VCH Weinsheim, **1998**; 138-144

14. Wais, A.; Witte, E.G.; De Graaf, A.A.; Mittelstaedt, W.; Haider, K.; Burauel, P.; Führ, F. In: *Pesticide Movement to Water*, Walker, A. et al., Eds., BCPC Monograph No 62, **1995**; 201-206

15. Wais, A.; Burauel, O.; De Graaf, A.A.; Haider, K.; Führ, F. *J. Environ.Sci. Health* B **1996**, 31, 1-25

16. Müller, L.; Mittelstaedt, W.; Pfitzner, J.; Führ, F.; Jarczyk, H.J. *Pestic. Biochem. Physiol.* **1983**; 19, 254-261

17. Kubiak, R.; Hansper, M.; Führ, F.; Mittelstaedt, W.; Steffens, W. In: *Factors Affecting Herbicidal Activity and Selectivity;* Proc. EWRS Symp., Wageningen, **1988**; pp 195-300 .

18. Kubiak, R.; Führ, F.; Mittelstaedt, W.; Hansper, M.; Steffens, W. *Weed Sci.* **1988**, 36, 514-518

19. Brumhard, B. Ph.-D. Thesis University of Bonn, **1991**

20. Pütz, T. Ph.-D. Thesis University of Bonn, **1993**

21. Printz, H. Ph.-D. Thesis University of Bonn, **1995**

22. Führ, F. In: *Pesticides Effects on Terrestrial Wildlife*; Somerville, L., Walter, C.H., Eds.; Taylor & Francis: London, 1990; pp 65-79
23. Printz, H.; Mittelstaedt, W.; Führ, F. *Environ. Sci. Health* B, 1995; 30, 269-288
24. Leake, C.R. *Pesticide Science* 1991, 31, 363-373
25. Leake, C.R.; Arnold, D.J.; Newby, S.E.; Somerville, L. Benazolin-ethyl - A case study of herbicide degradation and leaching. *British Crop Protection Conference-Weeds, British Crop Protection Council*, 1987, 577-583
26. Roberts, T. R.; Klein, W.; Still, G. G.; Kearney, P.C.; Drescher, N.; Desmoras, J.; Esser, H. O.; Aharonson, N; Vonk, J. W. Pure Appl. Chem., 1984, 56, 945-956
27. Calderbank, A. *Rev. Environ. Contam. Toxicol.* 1989, 72-103
28. Führ, F.; Ophoff, H.; Burauel, P.; Wanner, U.; Haider, K. In: *Pesticide Bound Residues in Soil.* Deutsche Forschungsgemeinschaft Senate Commission for the Assessment of Chemicals Used in Agriculture Report 2, Wiley-VCH Weinheim 1998, 175-176

Chapter 2

Improved Enzyme-Linked Immunosorbent Assay for the Insecticide Imidacloprid

Hee Joo Kim[1], Shangzhong Liu[1], Young Soo Keum[1], Eul Chul Hwang[2], and Qing X. Li[1]

[1]Department of Molecular Biosciences and Bioengineering, University of Hawaii, Honolulu, HI 96822
[2]Department of Applied Biology, Dong-A University, Pusan, Korea

Imidacloprid, 1-[(6-chloro-3-pyridinyl)methyl]-N-nitro-2-imidazolidinimine, is a systemic insecticide used worldwide. The sensitivity of an enzyme-linked immunosorbent assay (ELISA) for imidacloprid was considerably improved with a purified polyclonal antibody (designated as Ab-IIa) and a direct competitive assay format. The assay conditions were optimized for buffer concentrations and pH, solvents, and Tween 20 concentrations. Under the optimized conditions, the half-maximal inhibition concentration and the limit of detection were approximately 1 μg/L and 0.06 μg/L, respectively. This means a 35-fold improvement in the assay detectability compared with the assay previously reported (Li and Li, *J. Agric. Food Chem.*, 2000, *48*:3378-3382). The assay was very specific to imidacloprid and showed little cross-reactivity with other structural analogs. Computational analysis suggests that the antibody specificity primarily relate to the dihedral angles between the two rings, steric hindrance and electrostatic charges on the imidazolidinyl ring. Such information is useful for hapten design. The ELISA analysis of water samples fortified with imidacloprid showed a satisfactory correlation with the fortified levels. This assay can be a rapid and sensitive method for monitoring imidacloprid residues in the environment.

Introduction

Imidacloprid, 1-[(6-chloro-3-pyridinyl)methyl]-N-nitro-2-imidazolidinim-ine, is a systemic insecticide for effective control of sucking insects. Imidacloprid blocks the nicotinergic receptor in insects, causing death (1-3). Current methods for the analysis of imidacloprid residues in environmental and biological matrices include high performance liquid chromatography (HPLC), gas chromatography (GC), HPLC-mass spectrometry (HPLC-MS) and GC-MS. Imidacloprid is thermolabile and polar, and thus derivatization is required for GC and GC-MS determinations (4-11). Immunoassays are valuable methods for environmental monitoring of pesticides (12-13). Polyclonal antibody-based indirect competitive ELISAs (cELISAs) were reported for the analysis of imidacloprid (14). One assay is specific to imidacloprid with a half-maximal inhibition concentration (I_{50}) of 35 ng/mL. Another can measure both imidacloprid and its major metabolites although the limit of detection is relatively high. Monoclonal antibody-based ELISAs were also recently developed for the chloronicotinoid insecticides imidacloprid and acetamiprid (15). In this study, several assay parameters were optimized to improve the assay sensitivity and analysis time. Computer modeling of imidacloprid and its structural analogs was used to aid understanding of the antibody specificity.

Experimental

Reagents

All reagents were of analytical grade unless specified otherwise. Imidacloprid (96.9% purity) and its metabolite standards which are imidacloprid olefin (99%), 5-hydroxyimidacloprid (100%) and 6-chloronicotinic acid (99%) were obtained from Bayer Corp, Stillwell, KS. Chemicals purchased from Sigma (St. Louis, MO) were goat anti-rabbit IgG-horse radishperoxidase (IgG-HRP) (A-6154), ovalbumin (OVA) (A-5503), protein A (P-6031), HRP (P-6782), phosphate-citrate buffer capsules with sodium perborate (P-4922) and carbonate-bicarbonate buffer capsules (C-3401), and o-phenylenediamine (OPD) (P-9029). The ELISAs were carried out in 96-well polystyrene microplates (MaxiSorp F96, Nalge Nunc International, Denmark). The haptens, hapten-II–OVA and hapten-II–HRP conjugates were synthesized according to the procedure previously described (14). The antiserum (designated as Ab-IIa), previously described (14), was purified with an immunopure IgG purification kit (Pierce, Rockford, IL) according to the manufacture's instruction. Concentration of IgG in the final preparation was determined with a BSA assay (Pierce, Rockford, IL). The purified IgG in 0.02 M of phosphate-buffered saline (PBS) was stored at –20 °C.

Direct cELISA

A microtiter plate was coated with protein A (0.1 μg in 100 μL/well in 0.1 M carbonate–bicarbonate buffer, pH 9.6) overnight at 4 °C. The plate was washed with 0.02 M PBS containing 0.05% Tween 20 (PBST) for four times and subsequently coated with purified Ab-IIa (25 ng in 100 μL/well) for 12 h at 4 °C. After washing for five times, the plate was blocked with 1% BSA in PBST (200 μL/well) for 1 h at room temperature. After the plate was washed with PBST (5x) again, 50 μL of the standard or sample at various concentrations in PBST were added, followed by addition of an enzyme tracer (1/16,000 dilution in PBST, 50 μL/well). The plate was incubated for 30 min at 37 °C, then washed with PBST (5x). The substrate solution (1.0 mg/mL of OPD in 0.05 M citrate-phosphate with 0.03% sodium perborate, pH 5.0) was added (100 μL/well), and the enzymatic reaction was stopped with sulfuric acid (2 M, 50 μL/well) after 15-20 min at room temperature. The absorbance was read with a Vmax microplate reader at 490 nm (Molecular Devices, Sunnyvale, CA). Inhibition curves were analyzed by mathematical fitting of experimental data to a four-parameter logistic equation with Softmax software.

Determination of Cross-Reactivity

The ability of the antibody Ab-IIa to recognize imidacloprid analogs was assessed by performing direct cELISAs to determine their respective I_{50} values. Cross-reactivity (CR) was calculated as (I_{50} by imidacloprid / I_{50} by a test compound) x 100.

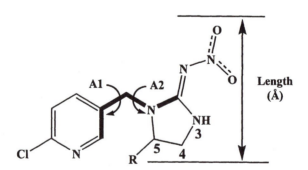

Figure 1. Length and dihedral angles of imidacloprid and its metabolites. Length is measured as the distance between the nitro oxygen and the far end atom of the R substituent. Dihedral angles (A-1 and A-2) are the angles between bold bonds.

Molecular Modeling

A global energy minimum structure was searched with HyperChem 5.0 and HyperSpin, an add-on program (Hypercube Inc., Gainesville, FL) running on an IBM personal computer. After the initial optimization with MM+ force field, low energy conformations were searched by sequentially rotating dihedral angles through 360° in 10° increments (Figure 1). The lowest energy conformation was selected from several local energy minimum structures through the optimization by a semi-empirical PM3 force field which was also used to calculate partial charge of each atom (16).

Electrostatic potentials of the global energy minimum structure were calculated with the CAChe Worksystem (Fujitsu, Beaverton, OR) using the PM3 force field (Plate 1). The results were visualized by the CAChe Tabulator application. Octanol -water partition coefficients were calculated from atomic partial charges with the CNDO force field (17) using Molecular Modeling Pro (ChemSW Software Inc., N. Fairfield, CA).

Fortification of Imidacloprid in Water

The optimized ELISA was used to determine imidacloprid concentrations in tap water samples fortified with imidacloprid up to 40 ng/mL. Final ionic strengths, as buffer concentrations, of the samples were adjusted with four-fold concentrated PBST.

Results and Discussion

Assay sensitivity and analysis time directly relate to assay formats and conditions. The objectives of this work were to improve the assay characteristics by optimizing assay conditions and using a direct cELISA, and to further confirm the high selectivity of Ab-IIa to imidacloprid. Computer modeling was used to propose structural geometry and electrostatic potential surface of different test molecules. I_{50} values of different chemicals to Ab-IIa were correlated with their electrostatic potential and geometry, in order to understand possible contributors of hapten features to the high antibody specificity.

Effect of Assay Buffer Ionic Strength

Large effects of ionic strength on immunoassays were often observed for many polar and nonpolar molecules (18-22). The assays were run in PBST varying concentrations from 0.02 M to 0.4 M and in deionized water as a control. All buffers contained 0.05% Tween-20 and had a pH of 7.5. Competitive imidacloprid inhibition curves generated at various buffer

34

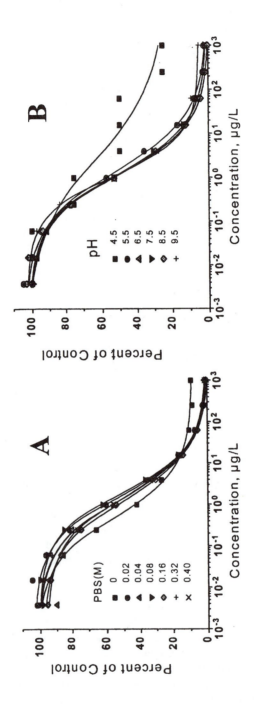

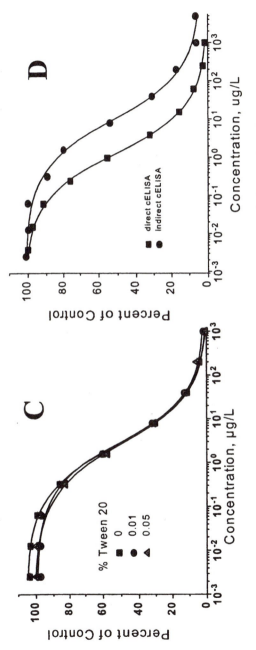

Figure 2. Effects of assay buffer concentration (**A**), pH (**B**) and Tween 20 (**C**) on the assay, and typical imidacloprid inhibition curves of direct and indirect cELISAs (**D**) at the optimized condition.

concentrations are shown in Figure 2A. There was no significant change in I_{50} values ($I_{50}s$) as the buffer concentration (*i.e.*, ionic strength) increased. Ionic strength slightly altered the inhibition curves only at low levels (0.06-1 µg/L) of imidacloprid (Figure 2A). The imidacloprid inhibition in deionized water was higher than that in PBST. Maximum absorbance of the assay was slightly higher in 0.02 and 0.04 M PBST than that in 4-20-fold-concentrated buffers and deionized water (data not shown).

pH Effect

Because the antigen-antibody binding occurs through weakly intermolecular interactions (23), and pH derived alteration of an analyte may lead to the poor recognition by an antibody (24), the evaluation of pH effect on the assay is a necessary step. Imidacloprid and the antibody were diluted in buffers of different pH values but with the same ionic strength. The buffer pH varying from 5.5 to 9.5 had negligible effect on inhibition curves (Figure 2B). The antibody activity was significantly reduced at pH 4.5. The assay signals (absorbance) also varied considerably as the pH was altered. Maximum signals were around pH 6.5-7.5. The results indicate that the assay is applicable in PBS with pH 5.5–9.5. However, it is noted that adjustment of buffer pH and concentrations (ionic strength) among samples is required to minimize errors due to their effect on absorbance. The buffer pH was kept at 7.0 for the subsequent work.

Effect of Tween 20

Tween 20 is a nonionic detergent commonly used in ELISAs to reduce non-specific interactions and often has strong effects on antibodies and assay characteristics (22, 25-26). In some circumstances, the assay detectability was considerably enhanced in detergent-free buffers (22, 25). Tween 20 (0.01-0.05%) showed little effect on this imidacloprid assay (Figure 2C). PBS containing 0.05% Tween-20 was used for the remainder of the study.

Typical Inhibition Curve

With the purified Ab-IIa, a direct cELISA format was employed to improve the assay characteristics since it was generally more sensitive than the indirect format. Figure 2D shows representative standard curves for imidacloprid obtained with direct and indirect cELISAs at the optimized assay conditions. The linear range of the direct and indirect cELISAs was approximately 0.15-15

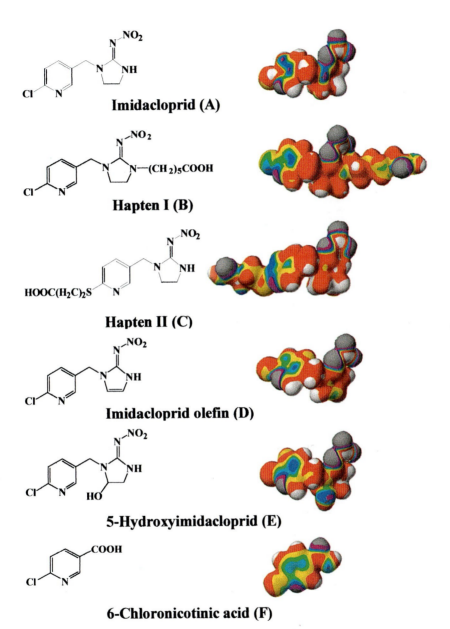

Imidacloprid (A)

Hapten I (B)

Hapten II (C)

Imidacloprid olefin (D)

5-Hydroxyimidacloprid (E)

6-Chloronicotinic acid (F)

Plate 1. Structures and electrostatic potentials mapped on electron density isosurfaces of imidacloprid and its analogs. The energy values in atomic unit on each surface color are: white, > +1; red, +0.09 to 0.03; yellow, +0.03 to +0.01; pale green, +0.01 to 0.00; pale blue, 0.00 to –0.01; blue –0.01 to –0.03; violet, -0.03 to –0.06; and black, < -0.06.

and 1-200 μg/L of imidacloprid, respectively. Concentrations giving 50% and 20% of inhibition with the direct cELISA were approximately 1.1 and 0.15 μg/L of imidacloprid, respectively, improving the sensitivity by over 35-fold compared with that of an indirect cELISA previously reported (14). Indirect cELISAs were tested with the purified Ab-IIa and hapten-II–OVA as a coating antigen to estimate the contribution of antibody purification to the assay improvement. The indirect cELISA gave competitive curves with an I_{50} of approximately 10 μg/L, which indicated that the antibody purification improved the assay sensitivity by 3- to 4-fold (14, and Figure 2D) when the same assay format and coating antigen were used. Antibody purification also reduced the assay background signal.

Solvent Effect

Antibody tolerance to solvents has been studied in many immunoassays (27-31). High solvent-tolerable ELISAs are very convenient and desirable for applications where sample extractions involve use of organic solvents. The effects of methanol (MeOH), acetone, acetonitrile, and dimethyl sulfoxide (DMSO) on the optimized assay were studied because they are commonly used in ELISA procedures. The solvent concentrations in the final assay varied from 0% to 40% in 0.02 M PBST. In general, all the solvents affected the assay signals. However, MeOH and DMSO have much smaller effects on the inhibition curves (shape and position) than acetone and acetonitrile (Figure 3).

Cross-Reactivity

Antibodies intrinsically recognize compounds structurally similar to the immunizing hapten (23). Five structural analogs of imidacloprid were examined for their cross-reactivity with the antibodies (Table I, Plate 1). Antiserum Ab-IIa was specific to imidacloprid and had low cross-reactivity (≤ 9%) for the metabolites and hapten-I (Table I). Ab-IIa highly recognized hapten-II, the immunizing hapten. It should be mentioned that Ab-IIa showed no recognition with other imidacloprid metabolites such as imidacloprid-guanidine and imidacloprid-urea, however, their purity was not confirmed (data not shown). The antibody specificity obtained with this direct cELISA agreed with that obtained with an indirect cELISA using purified Ab-IIa in this study (data not shown) and that previously determined with the indirect cELISA using the crude antiserum Ab-IIa (14).

Molecular Modeling of Imidacloprid Analogs

Three dimensional geometry, hydrophobicity and electrostatic properties are calculated in order to understand the selective binding of Ab-IIa to imidacloprid (Figure 1, Plate 1 and Table I). No recognition to 6-chloronicotinic acid suggests that the whole molecule or imidazolidinyl ring is the essential binding moiety. The large I_{50} differences between imidacloprid and the other

38

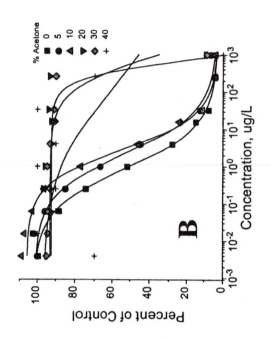

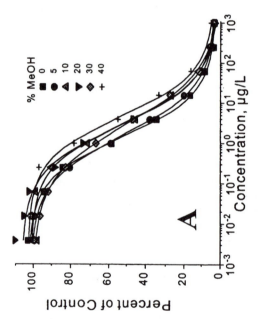

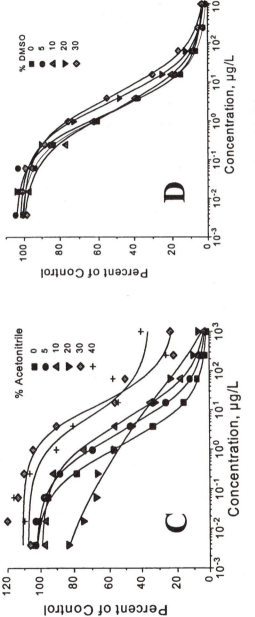

Figure 3. Effects of methanol (A), acetone (B), acetonitrile (C) and DMSO (D) on the assay.

analytes (**B**, **D-F** in Table I) indicate that 3-dimensional geometry of the whole molecule is important for the selective recognition. The molecular configuration between the imidazolidinyl and pyridinyl rings (*i.e.*, angles A1 and A2 in Figure 1) and the distance (designated as length) between the NO_2 oxygen and the end atom of the substituent R at the 5 position of the imidazolidinyl ring may play an important role in the antibody selectivity. The differences of the corresponding dihedral angles between imidacloprid (**A**) and hapten-II (**C**) (-0.22° and 1.64° for A1 and A2, respectively) are much smaller than those between **A** and hapten-I (**B**), imidacloprid olefin (**D**), or 5-hydroxyimidacloprid (**E**) (3.2-16° and 3.75-11.98° for A1 and A2, respectively). These computational results support the antibody selectivity, *i.e.*, low I_{50}s for **A** and **C**, high I_{50}s for **D** and **E** and no competition for **B** and **F** (Table I).

Table I. Comparison of Assay Cross-Reactivity and Length and Dihedral Angles of Imidacloprid and Its Structural Analogs

Chemical	I_{50}, ng/mL	CR,[a] %	Length,[b] Å	Dihedral angle[b] A-1	Dihedral angle[b] A-2
Imidacloprid (**A**)	1	100	6.525	70.03	66.59
Hapten-I (**B**)	nc[c]	0	6.525	73.23	70.34
Hapten-II (**C**)	0.7	143	6.526	69.81	68.23
Imidacloprid olefin (**D**)	11	9	6.755	86.03	86.57
5-Hydroxyimidacloprid (**E**)	47	2	7.485	75.24	86.31
6-Chloronicotinic acid (**F**)	nc[c]	0	–	–	–

[a] CR = cross reactivity.
[b] See Figure 1 for the length and dihedral angle.
[c] nc = no competition.

Steric hindrance is also important to antibody recognition. The lengths of **A**, **B** and **C** are virtually identical, and notably shorter than those of **D** and **E**. The lack of Ab-IIa's recognition of **E** suggests the strict steric requirement for the antibody binding. The length of **E** is approximately 0.96 Å longer than that of **A**, in which the hydroxyl group may cause steric exclusion in antibody binding. When compared with **A**, **B** has a long and bulky hexanoic acid linker (6.734 Å) that presumably prevents binding (Plate 1).

Electrostatic properties of ligands play an important role in protein-ligand interactions (32-36). The strong negative charges on the nitro oxygen (from − 0.567 to −0.679) and large dipole moment (-3.59 D) (37) may have a strong electrostatic interaction with amino acid residues (*e.g.*, asparagine) in the antibody binding pocket (32). Charge distribution of **D**, having a similar molecular size as **A**, largely differs from that of **A**, particularly on the imidazolidinyl ring (Plate 1). The partial charges of hydrogen atoms in the 4 and 5 positions of imidazolidinyl in **A** (0.10) are half of that (0.20) in **D**. The calculated charge difference between the carbon and hydrogen atoms correspondingly at the 4 and 5 positions also shows that the 4,5-double bond in **D** is more polarized than the single bond in **A**. The 4,5-double bond may also induce the hydrogen at the N3 position to be more positive. Negative potentials from the 4,5-double bond in **D** can cause it to be more hydrophilic. Octanol-water partition coefficients (K_{ow}) of **A** and **D** are 5.02 and 2.92, respectively, which suggests that the double bond increases the hydrophilicity of **D**. Introduction of a hydroxyl group at the 5 position also increases the hydrophilicity of **E**.

The electrostatic potential surfaces on pyridinyl rings in **A** and **C** are considerably different, but almost identical on imidazolidinyl rings. The I_{50} values are almost the same for **A** and **C**. Therefore, the imidazolidinyl ring may primarily govern the high antibody specificity, and the pyridinyl ring on which the hapten spacer is attached may affect the molecular geometry for proper binding. The results of this study support the usefulness of molecular modeling aiding rational designs of haptens and immunoassay development (38-40).

Analysis of Imidacloprid in Water

Tap water was fortified with imidacloprid at different levels up to 40 ng/mL and assayed with the optimized direct cELISA. The concentrations of imidacloprid determined by ELISA correlated very well with the fortification values with a slope of 1.0 and a correlation coefficient of 1.00 (Figure 4). The results showed that the ELISA could accurately measure the concentration of imidacloprid in water. The direct cELISA took a much shorter time than the indirect cELISA previously reported (14).

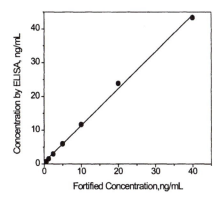

Figure 4. Correlation between concentrations of imidacloprid determined by ELISA and those fortified in tap water.

42

Conclusion

The detectability and analysis speed of an imidacloprid ELISA were significantly improved by purification of the polyclonal antibody, use of a direct cELISA format, and optimization of assay conditions. The assay was specific to imidacloprid with an I_{50} of approximately 1 ng/mL and had minimal cross reactivity with major imidacloprid metabolites. The antiserum (Ab-IIa) was raised from rabbits immunized with an immunogen of which the hapten had a spacer on the pyridinyl ring of imidacloprid. Computational analysis suggests that the antibody specificity primarily relate to the dihedral angles between the two rings, steric hindrance and electrostatic charges on the imidazolidinyl ring. The distance between the two extremities of the imidazolidinyl ring may be critical for proper binding. The strong negative charges of the nitro group and large dipole moment may also be important for imidacloprid binding. Low antibody cross-reactivity to the analytes having a double bond or a hydroxyl group on the imidazolidinyl ring further suggests that the imidazolidinyl moiety is critical for proper antibody binding.

Acknowledgement

This work was supported in part by the State of Hawaii Department of Agriculture (Pesticides Branch). We thank Bayer Corp. for providing the metabolite standards, and Judith R. Denery for reviewing this manuscript.

Literature Cited

1. Bai, D.; Lummis, S. C. R.; Leicht, W.; Breer, H.; Sattelle, D. B. Actions of imidacloprid and a related nitromethylene on cholinergic receptors of an identified insect motor neurone. *Pestic. Sci.* **1991**, *33*, 197-204.
2. Buckingham, S. D.; Lapied, B.; Le Corronc, H.; Grolleau, F.; Sattelle, D. B. Imidacloprid actions on insect neuronal acetylcholine receptors. *J. Experimental Biology* **1997**, *200*, 2685-2692.
3. Liu, M.-Y.; Casida, J. E. High affinity binding of [³H]imidacloprid in the insect acetylcholine receptor. *Pestic. Biochem. Physiol.* **1993**, *46*, 40-46.
4. Baskaran, S.; Kookana, R. S.; Naidu, R. Determination of the insecticide imidacloprid in water and soil using high-performance liquid chromatography. *J. Chromatogr. A* **1997**, *787*, 271-275.
5. Fernandez-Alba, A. R.; Valverde, A.; Agüera, A.; Contreras, M.; Chiron, S. Determination of imidacloprid in vegetables by high-performance liquid chromatography with diode-array detection. *J. Chromatogr. A* **1996**, *721*, 97-105.
6. Ishii, Y.; Kobori, I.; Araki, Y.; Kurogochi, S.; Iwaya, K.; Kagabu, S. HPLC determination of the new insecticide imidacloprid and its behavior in rice and cucumber. *J. Agric. Food Chem.* **1994**, *42*, 2917-2921.

7. Macke, M. Quantitation of imidacloprid in liquid and solid formulations by reversed-phase liquid chromatography: Collaborative study. *JAOAC International*. **1998**, *81*, 344-348.

8. Placke, F. -J.; Weber, E. Method of determining imidacloprid residues in plant materials. *Plfanzenschtz Nachr*. Bayer, **1993**, *46*, 109-182.

9. Vilchez, J. L.; El-Khattabi, R.; Fernández, J.; Gonález-Casado, A.; Navalón, A. Determination of imidacloprid in water and soil samples by gas chromatography-mass spectrometry. *J. Chromatogr. A* **1996**, *746*, 289-294.

10. MacDonald, L. M.; Meyer, T. R. Determination of imidacloprid and triadimefon in white pine by gas chromatography/mass spectrometry. *J. Agric. Food Chem*. **1998**, *46*, 3133-3138.

11. Rouchaud, J.; Gustin, F.; Wauters, A. Soil biodegradation and leaf transfer of insecticide imidacloprid applied in seed dressing in sugar beet crops. *Bull. Environ. Environ. Toxicol*. **1994**, *53*, 344-350.

12. U.S. EPA. Immunoassay methods. In *SW-846 Test Methods for Evaluating Solid Waste Physical/Chemical Methods*. Revision 4.; U.S. EPA, Office of Solid Waste: Washington, DC, **1996**; Chapter 4.

13. Hage, D. S. Immunoassays. *Anal. Chem*. **1999**, *71*, 294R-304R.

14. Li, K.; Li, Q. X. Development of an enzyme-linked immunosorbent assay for the insecticide imidacloprid. *J. Agric. Food Chem*. **2000**, *48*, 3378 - 3382.

15. Wanatabe, S.; Ito, S.; Kamata, Y.; Omoda, N.; Yamazaki, T.; Munakata, H.; Kaneko, T. Yuasa, Y. Development of competitive enzyme-linked immunosorbent assays (ELISAs) based on monoclonal antibodies for chloronicotinoid insecticides imidacloprid and acetamiprid. *Anal. Chim. Acta* **2001**, *427*: 211-219.

16. Zerner, M. C. Semiempirical molecular orbital methods. In: Reviews in computational chemistry; Vol. II. Lipkowitz, K. B.; Boyd, D. B. VCH Publishers, New York, **1991**, pp313-365.

17. Bersuker, I. B.; Dimoglo, A. S. The electron-topological approach to the QSAR problem. In: Reviews in computational chemistry; Vol. II. Lipkowitz, K. B.; Boyd, D. B. VCH Publishers, New York, **1991**, pp 423-460.

18. Manclús, J. J.; Montoya, A. Development of an enzyme-linked immuno-sorbent assay for 3,5,6-trichloro-2-pyridinol. 2. Assay optimization and application to environmental water samples. *J. Agric. Food Chem*. **1996**, *44*, 3710-3716.

19. Manclús, J. J.; Montoya, A. Development of enzyme-linked immunosorbent assays for the insecticide chlorpyrifos. 2. Assay optimization and application to environmental waters. *J. Agric. Food Chem*. **1996**, *44*, 4063-4070.

20. Abad, A.; Montoya, A. Development of an enzyme-linked immunosorbent assay to carbaryl. 2. Assay optimization and application to the analysis of water samples. *J. Agric. Food Chem*. **1997**, *45*, 1495 - 1501.

21. Abad, A.; Moreno, M. J.; Montoya, A. Development of monoclonal antibody-based immunoassays to the N-methylcarbamate pesticide carbofuran. *J. Agric. Food Chem.* **1999**, *47*, 2475 - 2485.

22. Galve, R.; Camps, F.; Sanchez-Baeza, F.; Marco, M.-P. Development of an immunochemical technique for the analysis of trichlorophenols using theoretical medels. *Anal. Chem.* **2000**, *72*, 2237 - 2246.

23. Padlan, E. A. Antibody-antigen complexes. R. G. Landes Co.: Austin, **1994**.

24. Schneider, P.; Gee, S. J.; Kreissig, S. B.; Harris, A. S.; Krämer, P.; Marco, M. P.; Lucas, A. D.; Hammock, B. D. Troubleshooting during the development and use of immunoassays for environmental monitoring. *In New Frontiers in Agrochemical Immunoassay*; Kurtz, D.A., Skerritt, J.H., Stanker, L., Eds.; AOAC International: Arlington, VA, **1995**; pp 103 - 122.

25. Shan, G.; Stoutamire, D. W.; Wengatz, I.; Gee, S. J.; Hammock, B. D. Development of an immunoassay for the pyrethroid insecticide esfenvalerate. *J. Agric. Food Chem.* **1999**, *47*, 2145 - 2155.

26. Chiu, Y.-W.; Chen, R.; Li, Q. X.; Karu, A. E. Derivation and properties of recombinant Fab antibodies to coplanar polychlorinated biphenyls. *J. Agric. Food Chem.* **2000**, *48*, 2614 - 2624.

27. Kido, H.; Goodrow, M.H.; Griffeth, V.; Lucas, A.D.; Gee, S.J.; Hammock, B.D. Development of an enzyme-linked immunosorbent assay for the detection of hydroxytriazines. *J. Agric. Food Chem.* **1997**, *45*, 414 - 424.

28. Shan, G.; Leeman, W. R.; Stoutamire, D. W.; Gee, S. J.; Chang, D. P. Y.; Hammock, B. D. Enzyme-linked immunosorbent assay for the pyrethroid permethrin. *J. Agric. Food Chem.* **2000**, *48*, 4032 – 4040.

29. Li, K.; Chen, R.; Zhao, B.; Liu, M.; Karu A. E.; Roberts, V. A.; Li, Q. X. Monoclonal antibody-based ELISAs for part-per-billion determination of polycyclic aromatic hydrocarbons: effects of haptens and formats on sensitivity and specificity. *Anal. Chem.* **1999**, *71*, 302 - 309.

30. Chiu, Y.-W.; Carlson, R. E.; Marcus, K. L.; Karu, A. E. A monoclonal immunoassay for the coplanar polychlorinated biphenyls. *Anal. Chem.* **1995**, *67*, 3829-3839.

31. Li, Q. X.; Hammock, B. D.; Seiber, J. N. Development of an enzyme-linked immunosorbent assay for the herbicide bentazon. *J. Agric. Food Chem.* **1991**, *39*, 1537 -1544.

32. Hsieh-Wilson, L. C.; Schultz, P. G.; Stevens, R. C. Insights into antibody catalysis: Structure of an oxygenation catalyst at 1.9 Å resolution. *Proc. Natl. Acad. Sci.* **1996**, *93*, 5363 – 5367.

33. Pellequer, J.-L.; Zhao, B.; Kao, H.-I; Bell, C. W.; Li, K.; Li, Q. X.; Karu, A. E.; Roberts, V. A. Stabilization of bound polycyclic aromatic hydrocarbons by a π-cation interaction. *J. Mol. Biol.* **2000**, *302*, 691-699.

34. Jeffrey, P. D.; Schildbach, J. F.; Chang, C. Y.; Kussie, P. H.; Margolies, M. N.; Sheriff, S. Structure and specificity of the anti-digoxin antibody 40-50. *J. Mol. Biol.* **1995**, *248*, 344-360.

35. Wedemayer, G. J.; Patten, P. A.; Wang, L. H.; Schultz, P. G.; Stevens, R. C. Structural insights into the evolution of an antibody combining site. *Science* **1997**, *276*, 1665 – 1669.

36. Heine, A.; Stura, E. A.; Yli-Kauhaluoma, J. T.; Gao, C. S.; Deng, Q.; Beno, B. R.; Houk, K. N.; Janda, K. D.; Wilson, I. A. An antibody exo Diels-Alderase inhibitor complex at 1.95 angstrom resolution. *Science* **1998**, *279*, 1934 – 1940.

37. Bowden, K., Electronic effects in drugs. *In Comprehensive medicinal chemistry*; Vol. 4. Ramsden, C. A. Vol. Ed., Pergamon Press, New York, **1990**, pp 205-239.

38. Holtzapple, C. K.; Buckley, S. A.; Stanker, L. H. Production and characterization of monoclonal antibodies against sarafloxacin and cross-reactivity studies of related fluoroquinolones. *J. Agric. Food Chem.* **1997**, *45*, 1984 -1990.

39. Muldoon, M. T.; Holtzapple, C. K.; Deshpande, S. S.; Beier, R. C.; Stanker, L. H. Development of a monoclonal antibody-based cELISA for the analysis of sulfadimethoxine. 1. Development and characterization of monoclonal antibodies and molecular modeling studies of antibody recognition. *J. Agric. Food Chem.* **2000**, *48*, 537 -544.

40. Wang, S.; Allan, R. D.; Skerritt, J. H.; Kennedy, I. R. Development of a class-specific competitive ELISA for the benzoylphenylurea insecticides. *J. Agric. Food Chem.* **1998**, *46*, 3330 -3338.

Chapter 3

Fate of the Insecticide Imidacloprid Treated to the Rice-Grown Lysimeters during Three Consecutive Cultivation Years

J. K. Lee[1], F. Führ[2], K. C. Ahn[1], J. W. Kwon[1], J. H. Park[1], and K. S. Kyung[3]

[1]Department of Agricultural Chemistry, Chungbuk National University, Cheongju 361–763, Korea
[2]Institute of Chemistry and Dynamics of the Geosphere IV: Agrosphere, Forschungszentrum Jülich GmbH, D–52425 Jülich, Germany
[3]National Institute of Agricultural Science and Technology, Pesticide Safety Division, RDA, Suwon 441–707, Korea

In order to elucidate the environmental fate of the insecticide imidacloprid being used in rice paddies in terms of leaching, volatilization and mineralization in soil, and absorption/translocation by rice plants, [imidazole-ring-2-^{14}C]imidacloprid (specific activity: 4.2 MBq/mg) was applied to two lysimeters (0.564 m ID×1 m soil depth) simulating the field conditions at the rate of 72 g a.i./ha, 28 days after rice transplanting. The amount of imidacloprid applied was three times larger than the ordinary dosage to ensure easy detection. Rice plants were grown by the conventional farming in Korea. The amount of $^{14}CO_2$ evolved from the flooded soil surfaces during the rice growing period (12 weeks), and that from the nonflooded soil surfaces for 9 weeks after harvest were 3.68–4.37% and 12.99–13.50% of the originally applied radioactivity, respectively, suggesting that the mineralization of imidacloprid in soil is much faster under the nonflooded aerobic conditions compared to flooded ones. The loss by volatilization from the soil was negligible as verified by sulfuric acid trapping. ^{14}C-Radioactivity detected from the leachates was 0.11–0.27% during the period of 131 weeks after application, indicating that imidacloprid and its degradation products move very slowly in soil. Average ^{14}C-radioactivity distributed in straw, ear without rice grains, chaff, and brown rice grains after harvest amounted to 1.80–1.94, 0.01, 0.05–0.06, and 0.07–0.09% of the originally applied radioactivity, respectively, in the first year, and 0.41–0.60, 0.002, 0.02, and 0.02, respectively, in the second year. The amounts decreased remarkably in all parts in the third year. 58–65% of the applied ^{14}C-radioactivity remained intact and/or degraded predominantly in the upper 15-cm layer of the soil after the second harvest (70 weeks after the application). Of the ^{14}C distributed in straw harvested in the first year, 88.36% was incorporated into the incompletely homogenized tissues and 4.05% into the nuclei.

Imidacloprid [1-(6-chloro-3-pyridinyl)methyl-4,5-dihydro-N-nitro-1H-imida-zole-2-amine] is a systemic insecticide belonging to the chemical group chloronicotinyl. It is known to act by binding to the nicotinergic acetylcholine receptor in the insect's nervous system (1,2) and to be effective against sucking insects, soil insects, termites, and some species of chewing insects. It is used as seed dressing, soil treatment, and foliar treatment in many crops (3). The high water solubility (610 mg/L at 20 ℃) (3) of imidacloprid indicates possible leachability in soil. However, sorption/desorption of pesticides in soil is an important factor affecting their leachability. Studies have reported low mobility of imidacloprid possibly due to high K_{oc} at low concentration (4) and an increase in sorption in soil with time (5,6) and with high contents of soil clay mineral and organic matters (7).

Hellpointner (8) reported a lysimeter study of imidacloprid after seed treatment of sugar beet followed by two crop rotations, showing the low mobility of imidacloprid and its degradation products in soil under practice-relevant field conditions.

Since imidacloprid is relatively new and in wide use in rice paddies in Korea, it is necessary to establish the standard for its safe use. For this purpose, lysimeters which can simulate our rice paddies were used for the conventional rice cultivation and the application of [^{14}C]imidacloprid, whereby the leaching, mineralization, volatilization, and bioavailability to rice plants could be elucidated under our environmental conditions.

Materials and Methods

Lysimeter and Preparation of Soil Cores

Two cylindrical lysimeters were manufactured with stainless steel (8 mm in thickness). Their surface area was 0.25 m^2, the height 1.1 m, and the inner diameter 0.564 m. The undisturbed soil cores were obtained by pressing down the lysimeters on a rice paddy soil located at Bockdae-dong, Cheongju, Korea, with the aid of a fork-crane. They were moved to the lysimeter station on campus and installed on the ground to collect the leachates through the soil

Table I. Physicochemical Properties of Soil Layers in the Lysimeters

Soil depth (cm)	pH H_2O (1:5)	O.M.[a] (%)	CEC^b (cmol$^+$kg^{-1} soil)	Sand	Silt	Clay	Texture
				---------------% -------------			
0–10	6.4	2.3	12.8	42.4	42.4	15.2	L^c
10–20	5.9	2.5	13.1	40.0	45.8	14.2	L
20–30	6.3	1.9	12.7	39.6	44.4	16.0	L
30–40	7.2	1.1	12.2	27.4	50.1	22.5	SiL^d
40–50	7.4	0.8	9.2	26.4	49.5	24.1	L
50–60	7.4	0.8	13.0	17.4	52.8	29.8	SiCL^e
60–70	7.2	0.8	14.1	28.6	47.0	24.4	L
70–80	6.9	0.6	11.9	38.0	41.0	21.0	L
80–90	6.9	0.5	9.6	36.6	44.3	19.1	L
90–100	6.5	0.5	11.1	28.0	51.3	20.7	SiL

[a]Organic matter, [b]Cation exchange capacity, [c]Loam, [d]Silt loam, [e]Silty clay loam

profile and $^{14}CO_2$ evolved from the soil surface. The physicochemical properties of the lysimeter soil layers are presented in Table I.

Growing of Rice Plants

Before transplanting, the lysimeter soils were fertilized with N-P-K at the ratio of 150-90-110 kg/ha, except that 80% of the total nitrogen fertilizer was applied at the beginning and the remaining 20% at the earing stage.

The rice seedlings (*Oryza sativa* cv. Akibare, Japan) were transplanted onto lysimeter soils in May or June. Twenty-seven seedlings (3 seedlings/hill×9 hills) were grown on each lysimeter soil. Throughout the cultivation period, the soils were flooded to simulate the real rice paddy fields. Accordingly, during the dry season, tap water was used as additional irrigation. After rice harvest, the lysimeters were kept open until the following cultivation. For the next cultivation, the upper 0–10 cm layer of the soil was mixed evenly and rice plants were grown in the same way.

The average monthly atmospheric and soil (20-cm depth) temperatures, and precipitation during the period of 1997–1999 are presented in Table II and Table III.

Table II. The Monthly Precipitation (mm) in Cheongju in 1997–1999

Year	Month											
	1	2	3	4	5	6	7	8	9	10	11	12
1997	-	-	-	-	-	-	425.5	211.1	55.5	8.4	180.3	44.3
1998	22.0	28.9	30.9	153.1	92.8	247.0	253.0	460.6	225.9	74.2	44.7	7.1
1999	1.6	3.6	54.1	91.4	102.4	191.1	122.4	197.4	281.3	252.4	15.4	13.4

Table III. The Average Monthly Atmospheric and Soil (20-cm depth) Temperatures ($^\circ C$) in Cheongju in 1997–1999

Year	Average temperature	Month											
		1	2	3	4	5	6	7	8	9	10	11	12
1997	Atmospheric	-	-	-	-	-	-	30.1	30.3	26.2	19.3	9.4	1.7
	Soil	-	-	-	-	-	-	28.1	27.2	22.5	15.3	8.2	0.2
1998	Atmospheric	-1.3	3.1	7.2	19.4	23.8	25.4	30.2	28.9	28.2	21.8	5.6	4.2
	Soil	0.2	1.2	5.1	18.4	22.1	24.3	27.0	27.0	24.8	18.8	6.4	1.9
1999	Atmospheric	2.7	1.9	10.2	20.3	24.0	28.8	30.4	30.6	28.7	23.4	21.2	6.8
	Soil	1.4	-0.5	8.3	18.9	21.9	25.5	26.8	27.4	24.3	18.5	15.3	4.8

Application of [^{14}C]Imidacloprid and Cultivation

The position labelled with ^{14}C in the chemical structure of imidacloprid purchased from International Isotopes Münich was the second carbon of imidazole ring. The specific activity was 4.2 MBq/mg and the purity greater than 99%.

The commercial wettable powder formulation, Konidan (1.6% imidacloprid + 7% tebufenozide + 91.4% adjuvant, etc.) in which imidacloprid was replaced by [^{14}C]imidacloprid as a tracer was applied onto lysimeter soils 28 days after transplanting. To ensure a high concentration for studying its long-term fate such as leaching and degradation in soil, and absorption/translocation by rice plants, the amount of imidacloprid applied onto the lysimeter soils was three times larger than the recommended rate of 24 g a.i./ha. in Korea.

The total amounts of active ingredient and radioactivity of imidacloprid applied were 1.81 mg and 7.61 MBq, respectively, in each lysimeter. The ^{14}C-labelled imidacloprid dissolved in about 10 mL of methanol and the formulation prepared without imidacloprid were added to about 200 g soil. After the methanol was completely evaporated, the treated soil was uniformly spread on the flooded rice-grown lysimeter soils. Rice plants were cultivated for three consecutive years according to the same method as described above.

Leachates

The leachates percolated through the lysimeter soils were collected in 2-L plastic containers, and the volume and radioactivity were measured weekly, if necessary. The radioactivity was measured by a liquid scintillation counter (LSC: Tri-Carb 1500, Packard Instrument Co., Downers Grove, Ill., U.S.A.) with automatic quench correction.

Mineralization and Volatilization of Imidacloprid

^{14}CO$_2$ and volatile substances evolved from the [^{14}C]imidacloprid-treated soil surface in the first year (for 21 weeks after the application) were trapped in 1 N NaOH and 0.1 N H$_2$SO$_4$, respectively, by the method of Lee et al. (9), and their radioactivity was measured by LSC biweekly.

Soil and Plant Sampling and their Radioactivity

After harvest, the soil sample of each 5-cm layer was collected down to the 40-cm depth in the first year (27 weeks after [^{14}C]imidacloprid application) and the 50-cm depth in the second year (70 weeks after the application), with a soil core sampler attached to a stainless steel core of 5.05-cm diameter and 100 cm^3 volume. The samples were taken vertically from three random spots of the soil, and those of each layer were combined together, air-dried, and ground in a mortar for analysis. The rice plants harvested were separated into straw, ear without grains, chaff, and brown rice grains, freeze-dried, and pulverized with a cutting mill for the radioactivity measurement by combustion.

Each plant or soil sample (0.3 g) was combusted with the Biological Oxidizer, OX-400 (R. J. Harvey Instrument Corporation, U.S.A.) to give $^{14}CO_2$ which was absorbed in the ^{14}C-cocktail (CARBO MAX PLUS LUMAC*LSC B. V., the Netherlands) for the radioactivity measurement. The toluene cocktail was used for ^{14}C-radioactivity dissolved in organic solvents that were evaporated before cocktail addition. The radioactivities of $^{14}CO_2$ trapped in 1 N NaOH and volatile substances absorbed in 0.1 N H_2SO_4 were measured using Aquasol (Du Pont, NEN Research Products, U.S.A.) as a liquid scintillation cocktail.

Microbial Activity as Measured by Dehydrogenase and Dimethyl Sulfoxide (DMSO) Reduction to Dimethyl Sulfide (DMS), and the Number of Microbial Colonies in Lysimeter Soils

For comparison of the metabolic activity of microorganisms inhabiting the lysimeter soil, before and after the growing of rice plants, the dehydrogenase activity of the soil was measured by the methods of Lee et al. (9) and Casida (10).

The DMSO reduction rate as a valuable parameter for determining microbial activities in soil was measured by the method of Alef and Kleiner (11).

For counting the number of microorganisms in soil, 1 g of soil (on a dry weight basis) was added to 9 mL of sterile distilled water and shaken for 2 h. The soil suspension was diluted adequately, 0.1 mL of which was spread on the nutrient broth medium and incubated at 30°C for 48 h. The colonies were counted with a colony counter (12).

Extractable and Nonextractable ^{14}C-Soil-Bound Residues

Each 30 g of air-dried soil collected from every 5-cm layer of the lysimeters was exhaustively extracted with methanol. All extracts were combined and the radioactivity was measured. Methanol turned out to be the best among the tested solvents to extract imidacloprid residues in soil. That is, the extraction efficiency of imidacloprid with methanol immediately after the treatment to soil was over 99%. The extracted soil was then combusted to determine the nonextractable soil-bound residues.

Partition of ¹⁴C-Radioactivity of Soil Extracts between Aqueous and Organic Phase

To partition the ^{14}C-radioactivity of the methanol extracts from each soil layer between aqueous and organic phase, 10 mL of each extract was evaporated by a bubbling air stream, added with 5 mL of distilled water, and mixed homogeneously. Five mL of methylene chloride were then added to it and the mixture was shaken vigorously. The radioactivity in the organic and aqueous phase was measured.

Partition of ¹⁴C-Radioactivity in Leachates between Aqueous and Organic Phase

The leachates collected from lysimeter I and II were partitioned between methylene chloride and distilled water to examine the changes in their polarities. That is, 20 mL of each leachate was shaken with 20 mL of methylene chloride in a separatory funnel, and the ^{14}C-radioactivities of the aqueous phase (5 mL) and organic phase (10 mL), respectively, were measured. The average values obtained from the two lysimeters were presented.

Distribution of ¹⁴C-Radioactivity in Subcellular Particles of Rice Straw

Sixty grams of sliced and freeze-dried rice straw after harvest were added to 350 mL of a buffer (13) consisting of 250 mM sucrose, 2 mM dithiothreitol, 0.5 mM PMSF, 0.1% BSA, and 25 mM Tris-Mes (pH 7.0), macerated, and smashed in a Waring Blendor. The slurry was homogenized more with an IKA®-ULTRA-TURRAX T25 at a speed of 24,000 rpm for 20 sec, three times. The homogenate was filtered through four-fold gauze and the incompletely homogenized tissues were separated. The filtrate was centrifuged at $500 \times g$ for 10 min to separate the supernatant and the sediment (nuclei). The resulting supernatant was centrifuged at $2,000 \times g$ for 10 min to obtain the supernatant and the sediment (chloroplasts). The supernatant was centrifuged again at $9,940 \times g$ for 20 min to get the supernatant and the sediment (mitochondria). Finally, the supernatant was centrifuged at $80,000 \times g$ for 60 min (Beckman SW-28 rotor) to separate the supernatant (cytosol and some ribosomes) and the sediment (microsomes). The incompletely homogenized tissues (cellulose *et al.*) were freeze-dried and combusted to measure ^{14}C-radioactivity and the other fractions were suspended in 5 mL of the buffer, 2 mL of which was measured for ^{14}C-radioactivity.

Results and Discussion

Leaching

Table IV presents the total amounts of leachates and ^{14}C-radioactivity collected from the two lysimeters over three consecutive years. It shows that the ^{14}C-radioactivity leached with 103.4−146.4 L of leachates amounted to 0.11–0.27% (1.99×10^{-3}–4.89×10^{-3} mg imidacloprid equivalent) of the originally applied amount (1.81 mg). Oi *(6)* investigated the time-dependent sorption of imidacloprid with two German soils, indicating that sorption of imidacloprid increased with residence time in soil and made it more resistant to leaching.

Cox *et al. (14)* reported that the sorption of imidacloprid with aging in three soils was highest in the soil with the highest organic carbon content, and increased by an average factor of 2.8 during the incubation period. They also reported that soil clay mineral and organic components had been shown to be responsible for sorption of many pesticides *(7)*.

In this context, the relatively high amount of organic matter (2.3−2.5% down to the 20-cm depth) in the lysimeter soils would be responsible for the slow movement of imidacloprid in this investigation.

Table IV. Amounts of Leachates and ^{14}C-Radioactivity Leached from the Rice-Grown Lysimeters Treated with [^{14}C]Imidacloprid

Period (Week)	Lysi-meter	Amounts of leachates (L)	[14]C leached (%)	Amount of imidacloprid (mg)	
				Applied	Leached (Imidacloprid equivalent[a])
1st year (1st–5th week)	I	11.7	0.02	1.81	0.36×10^{-3}
	II	17.8	0.01	"	0.18×10^{-3}
1st–2nd year (1st–77th week)	I	45.8	0.04	"	0.72×10^{-3}
	II	53.5	0.03	"	0.54×10^{-3}
1st–3rd year (1st–131st week)	I	103.4	0.11	"	1.99×10^{-3}
	II	146.4	0.27	"	4.89×10^{-3}

[a]Calculated on the basis of the specific ^{14}C-radioactivity of imidacloprid applied

Mineralization and Volatilization

As shown in Figure 1, the amounts of $^{14}CO_2$ evolved from the flooded soil surfaces of lysimeter I and II during the cultivation period of 12 weeks after the application of [^{14}C]imidacloprid were 3.68 and 4.37%, respectively, of the originally applied ^{14}C-radioactivity. Meanwhile, the amounts of $^{14}CO_2$ evolved from the dry soil surfaces for 9 weeks after harvest increased remarkably up to 12.99 and 13.50% in lysimeter I and II, respectively.

This result strongly indicates that the microbial degradation of imidacloprid is predominant in soil and furthermore it is more favored under the aerobic conditions than the flooded anaerobic ones. In addition, the abrupt increase in $^{14}CO_2$ evolution will be due to the fact that many intermediate degradation products formed photochemically and microbiologically during the cultivation period were able to be readily mineralized to $^{14}CO_2$ under the dry aerobic conditions. Volatilization of the chemical from the lysimeter soil surfaces during the whole period was negligible.

Distribution of ^{14}C-Radioactivity in Different Parts of Rice Plants

The ^{14}C-radioactivity distributed in different parts of rice plants after harvest is presented in Table V. It was detected in all parts such as straw, ear without rice grains, chaff, and brown rice grains until the third cultivation year, decreasing remarkably year after year. The total amounts (%) absorbed and translocated by rice plants except for roots throughout the three years were in the range of 2.51–2.86% of the originally applied ^{14}C. Even if more ^{14}C-radioactivity was detected in straw than in any other parts, it decreased remarkably in the following years. In terms of risk assessment, the residual imidacloprid equivalent detected in brown rice grains, as calculated on the basis of the specific ^{14}C-radioactivity, was far below the MRL (maximum residue limit) value of 0.05 ppm set for rice by Korea Food & Drug Administration.

Distribution of ^{14}C-Radioactivity in Soil Layers

Figure 2 shows the distribution of ^{14}C-radioactivity in different soil layers of the two lysimeters after the first and second year of the ^{14}C-imidacloprid application and the subsequent cultivation of rice plants. Most of the applied ^{14}C-radioactivity (79.29–82.30 %) remained in the 0–10 cm soil depth in the first year (27 weeks after [^{14}C]imidacloprid application), but it decreased remarkably (62.62–53.92% in the same layer) in the second year (70 weeks after the application), due to the downward movement and degradation. However, as the data in Figure 3 demonstrate, only small amounts of ^{14}C-labelled compounds moved below the 0–30 cm plow layer.

The total ^{14}C-radioactivities remaining down to the 50-cm depth of the lysimeter soils after one year were 81.41 and 86.98 % of the originally applied

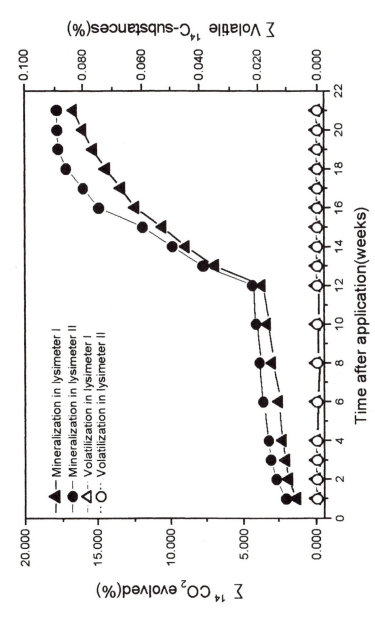

Figure 1. Mineralization of [^{14}C]imidacloprid to $^{14}CO_2$ and volatilization during the period of 21 weeks in the first year. ^{14}C-Radioactivity applied=100%. 1st–12th week: flooded rice-growing period, 13th–21st week: nonflooded dry period after harvest.

amounts in lysimeter I and II, respectively. Meanwhile, in the second year those were 66.89 and 61.49% in the same layers of the two lysimeter soils, respectively, indicating that some kinds of degradation leading to $^{14}CO_2$ evolution had occurred with time.

Microbial Activity in Lysimeter Soils

Both dehydrogenase activity and DMSO reduction are used as an index for the metabolic activity of microorganisms in soil. Casida *(10)* used a modified technique for measuring the dehydrogenase activity. Alef and Kleiner *(11)* presented a method for determining the microbial activity in soils and soil aggregates, based on the enzymatic reduction of DMSO to DMS, showing that 95% of about 150 tested strains of microorganisms were able to do this.

It was reported that no relationship was found between dehydrogenase activity and numbers of aerobic bacteria in rice field soils *(15)* and in corn field soil *(16)*. Nevertheless, as seen in Table VI, both the dehydrogenase activity as shown by the amount of formazan formed via various electron-donating substrates and the DMSO reduction as shown by the amount of DMS increased after rice cultivation. This result indicates that the degradation of imidacloprid and the formation of bound residues of imidacloprid and its degradation products would have been affected by the increased microbial activity by virtue of the rhizosphere effects of the rice cultivation.

Extractable and Nonextractable ^{14}C-Soil-Bound Residues and Partition of the Extractable ^{14}C between Aqueous and Organic Phase after the First Harvest

About 80–91% of the applied $[^{14}C]$imidacloprid was strongly adsorbed in the upper 0–10 cm soil depth where larger amounts of organic matter are contained in the form of nonextractable bound residues after the first harvest (Table VII). It suggests that imidacloprid and its degradation products in soil were tightly bound to soil organic matter and soil clay mineral as the nonextractable bound residues, so they could not be extracted by organic solvents. Oi *(6)* and Cox *et al.* *(4,7,14)* reported that the sorption of imidacloprid and its products by the chemical or microbial degradation was dependent on organic matter, clay mineral, and residence time in soil, and the degradation products with polar functional groups were more adsorptive to soil than imidacloprid.

As seen in Table VIII, the more ^{14}C partitioned in the aqueous phase from the 0–5 cm depth than that from the 5–10 cm indicates that the photochemical and microbiological degradation of imidacloprid leading to more polar degradation products was more favored in the upper 0–5 cm depth of the soil.

The increase in nonextractable bound residues in the 0–5 cm depth would be related to the increased polarity of the extract in that layer.

Table V. Amounts (%) of ^{14}C-Radioactivity Remaining in Each Part of Rice Plants Grown on the Lysimeter Soils Treated with [^{14}C]Imidacloprid. ^{14}C-Radioactivity Applied =100%

Parts of rice plants	Lysimeter I			Lysimeter II			MRL[c] (μg/g)
	1st year	2nd year	3rd year	1st year	2nd year	3rd year	
Straw	1.795±0.220 (0.150±0.018)[a]	0.405±0.006 (0.019±0.000)	0.128±0.010 (0.007±0.001)	1.935±0.203 (0.166±0.017)	0.598±0.020 (0.025±0.001)	0.136±0.013 (0.007±0.001)	
Ears without rice grain	0.008±0.001 (0.019±0.001)	0.002±0.000 (0.005±0.000)	0.002±0.000 (0.003±0.000)	0.008±0.000 (0.015±0.001)	0.002±0.000 (0.005±0.000)	0.002±0.000 (0.003±0.000)	
Chaff	0.056±0.002 (0.020±0.001)	0.019±0.000 (0.007±0.000)	0.010±0.000 (0.002±0.000)	0.051±0.002 (0.014±0.000)	0.015±0.001 (0.006±0.000)	0.010±0.001 (0.004±0.000)	
Brown rice grain	0.065±0.005 (0.006±0.000)	0.018±0.001 (0.002±0.000)	0.001±0.000 (0.001±0.000)	0.088±0.007 (0.007±0.001)	0.018±0.001 (0.002±0.000)	ND[b] ND	0.05
Total	1.924	0.444	0.141	2.082	0.633	0.142	

[a]Figures in parentheses represent concentrations of imidacloprid equivalent (μg/g) calculated on the basis of the specific ^{14}C-radioactivity (4.20 MBq/mg) of the imidacloprid applied. [b]Not detected, [c]Maximum residue limit, [d]Figures represent mean±standard deviation of triplicates.

58

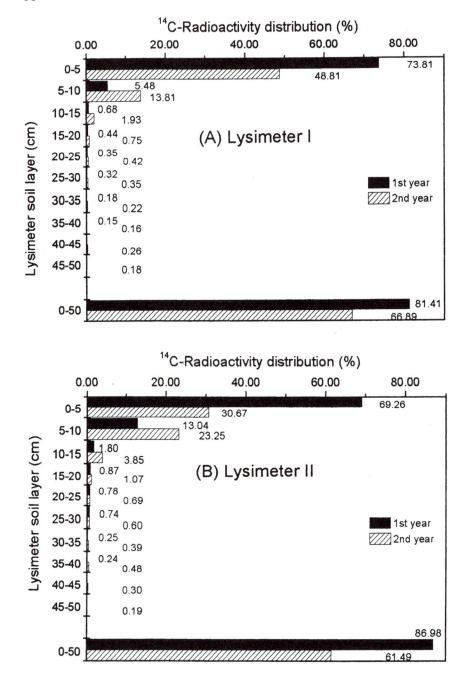

Figure 2. Amounts (%) of the ^{14}C-radioactivity distributed in the different layers of lysimeter soils treated with $[^{14}C]$imidacloprid after harvest. ^{14}C-Radioactivity applied=100%.

Table VII. Comparison of ^{14}C-Radioactivity Extracted and Nonextracted from the Lysimeter Soils Treated with [^{14}C]Imidacloprid and Collected after Harvest in the First Year

Lysimeter	Soil depth (cm)	^{14}C extracted with methanol	Nonextractable ^{14}C-bound residues
		---------------- % ------------------	
I	0–5	14.07	85.93
	5–10	19.92	80.08
II	0–5	8.56	91.44
	5–10	11.32	88.68

Table VI. Comparison of the Number of Microbial Colonies and the icrobial Activities in Terms of Dehydrogenase Activity and DMSO Reduction in the Topsoil (0-10 cm) of the Lysimeter I before and after the Cultivation of Rice Plants in the First Year

Vegetation	Number of colonies ($\times 10^6$ CFU^a/ g soil^b)	Formazan formed (mg/5 g soil)				DMS formed (ng DMS/ g soil $\times$ hour)
		Control	Electron-donating substrate			
			Glucose	Yeast extract	Brain heart infusion broth	
Before	6.7	0.097	0.086	0.164	0.311	11.970
After	13.5	0.126	0.089	0.230	0.375	14.740

^{a}Colony forming unit, ^{b}Dry weight basis

Table VIII. Distribution of ^{14}C-Radioactivity in the Methanol Extracts from the Different Layers of the Lysimeter Soils Treated with [^{14}C]Imidacloprid after Harvest between Aqueous Phase and Organic Phase. Aqueous Phase + Organic Phase = 100%

Lysi-meter	Soil depth (cm)	Extracted with methanol (%)	Distribution (%) of ^{14}C after partitioning	
			Aqueous phase	Organic phase (CH$_2$Cl$_2$)
I	0–5	14.07	14.63	85.37
	5–10	19.92	2.63	97.37
II	0–5	8.56	20.92	79.08
	5–10	11.32	15.41	84.59

Changes in the Polarity of the ^{14}C in Leachates

When each leachate collected at intervals from the lysimeter I and II was partitioned with methylene chloride to examine how much of the imidacloprid applied to the lysimeter soils was transformed into polar metabolites, the ^{14}C-radioactivity partitioned into the aqueous phase ranged from 79.78 to 94.68 % as seen in Figure 3.

Even if imidacloprid has a relatively high water solubility, most of the intact imidacloprid would be partitioned into the organic phase of methylene chloride. Taking account of this fact, more than 80% of imidacloprid applied must have been transformed into more polar metabolites in the course of leaching, and hence the possibility of groundwater contamination by intact imidacloprid could be ruled out.

Distribution of ^{14}C-Radioactivity in Subcellular Particles of Rice Straw

Because more ^{14}C-radioactivity was detected in rice straw than any other parts of rice plants, the ^{14}C-radioactivity distributed in each fraction of rice straw was measured. As seen in Table IX, 88.4% of the ^{14}C detected in rice straw was distributed in the incompletely homogenized tissues (cellulose et al.) This gives evidence that at least part of the ^{14}C detected in the straw was assimilated as ^{14}CO$_2$, the end product of [^{14}C]imidacloprid mineralization. The parent compound and its degradation products preferentially were adsorbed onto cellulose or other organics found in the debris.

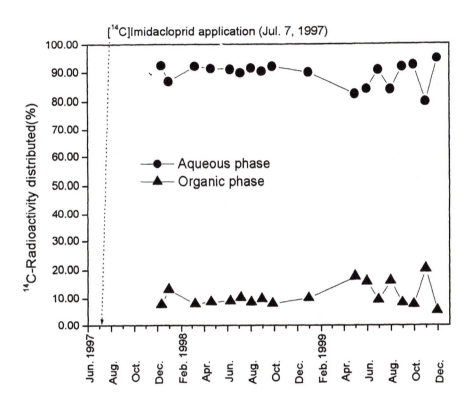

Figure 3. Changes in the distribution of ^{14}C-radioactivity in the leachates from lysimeter I and II between aqueous and organic phase with time. The values are the mean of those obtained from the two lysimeters.

Table IX. Distribution of ^{14}C-Radioactivity in Subcellular Particles of Rice Straw

Subcellular fraction	^{14}C-Radioactivity distributed (%)
Tissue debris (Incompletely homogenized tissues)	88.36
Mainly nuclei (500×g, 10 min)	4.05
Chloroplasts (2,000×g, 10 min)	2.33
Mitochondria (9,940×g, 20 min)	1.80
Microsomes (80,000×g, 60 min)	0.99
Solubles (Cytosol + some ribosomes; 80,000×g, 60 min)	2.47

Fate of [^{14}C]Imidacloprid in Rice Plant-Grown Lysimeters

Table X summarizes the fate of [^{14}C]imidacloprid treated onto the rice plant-grown lysimeter soils over the three consecutive years of cultivation. The total ^{14}C absorbed and translocated by rice plants cultivated during this period amounts to 2.50–2.85% of the originally applied amount. Moreover, the total ^{14}C detected in the leachates throughout three years (131 weeks after the application) is 0.11–0.27%. In soil, most of the imidacloprid applied remained mainly in the upper 0–10 cm layer and was degraded photochemically and microbiologically to polar products and to [14]CO_2, as seen in the leachates (Figure 3 and 4). The study on the time-dependent sorption of imidacloprid would explain its small leaching in soil (5,6).

The decrease in ^{14}C-radioactivity in the second year would be due to the loss by [14]CO_2 evolution indicated by the letter α in Table X. The final balance of ^{14}C-radioactivity will be established after harvesting the fifth rice crop which will be planted in 2002.

Considering the fact that the original amount applied was three times larger than the ordinary dosage, the real ^{14}C-radioactivity to be detected in the samples would be far less than these values presented here.

Table X. Fate of [¹⁴C]Imidacloprid Treated onto Rice Plant-Grown Lysimeter Soils during the Experimental Period. ¹⁴C-Radioactivity Applied = 100%

Period of cultivation	Lysimeter	$^{14}CO_2$ evolved	^{14}C volatile	^{14}C in rice plants	^{14}C leached	^{14}C in soil	Recovery
				------- % -------			
1st year	I	16.67	BG[a]	1.92	0.02	81.41[c]	100.02
	II	17.87	BG	2.08	0.01	86.98	106.94
1st – 2nd year	I	16.67+α[b]	BG	2.36	0.04	66.89[d]	85.96
	II	17.87+α	BG	2.71	0.03	61.49	82.10
1st – 3rd year	I	16.67+α	BG	2.50	0.11	-[e]	-
	II	17.87+α	BG	2.85	0.27	-	-

[a]Background, [b]Loss by ¹⁴CO₂ evolution not measured in the second and third cultivation years, [c]Down to 0–40 cm soil layer, [d]Down to 0–50 cm soil layer, [e]To be measured in the fifth year

Conclusions

Ever since imidacloprid was registered as an insecticide for protecting many crops including rice in Korea, it has been in wide use. Considering its relatively high water solubility, the concern about the possible imidacloprid contamination of groundwater and rice can not be ruled out.

Nevertheless, the results obtained from this investigation that most of the ¹⁴C-imidacloprid applied to the two lysimeter soils remained in the upper 0–10 cm layer during the two-year period, decreasing with time, and the polarity of ¹⁴C in the leachates increased remarkably indicate its safety in terms of groundwater protection.

Imidacloprid absorbed and translocated from soil to rice straw might be incorporated mainly into cellulose and some other constituents in the incompletely homogenized tissues and be rarely detectable in the microsomal and cytosolic fractions where hydrolytic enzymes responsible for its metabolism exist.

Moreover, the ¹⁴C detected in various parts of rice plants indicates a very little possibility of rice contamination, as verified by the rapid decrease of the chemical with time. Especially, the ¹⁴C corresponding to the imidacloprid equivalent detected in the edible brown rice grain is only one seventh to one eighth of the maximum residue limit (MRL) set by Korea Food & Drug Administration.

64

This lysimeter study suggests that the use of imidacloprid in rice paddies would not pose any problem with the contamination of groundwater and the harvested products, if the good agricultural practice (GAP) is observed carefully by the users.

Acknowledgments

The authors greatly appreciate the financial support of the Korea Science & Engineering Foundation and Deutsche Forschungsgemeinschaft.

References

1. Bai, D.; Lummis, S. C. R.; Leicht, W.; Breer, H.; Sattelle, B. D. *Pestic. Sci.* **1991**, *33*, 197-204.
2. Mullins, J. W. In *Pest Control with Enhanced Environmental Safety*, ACS Symposium Series 524; American Chemical Society: Washington, DC, **1993**; pp 183-198.
3. *The Pesticide Manual*; 11th ed.; C. D. S. Tomlin ed.; British Crop Protection Council, **1997**; p 706.
4. Cox, L.; Koskinen, W. C.; Yen, P. Y. *J. Agric. Food Chem.* **1997**, *45*, 1468-1472.
5. Walker, A.; Welch, S. J.; Turner, I. J. In *BCPC Monograph 62, Pesticide Movement to Water*; BCPC; Bracknell, U.K., **1995**; pp 13-24.
6. Oi, M. *J. Agric. Food Chem.* **1999**, *47*, 327-332.
7. Cox, L.; Koskinen, W. C.; Celis, R.; Yen, P. Y.; Hermosin, M. C.; Cornejo, J. *Soil Sci. Soc. Am. J.* **1998**, *62*, 911-915.
8. Hellpointner, E. In *Lysimeter Concept. The Environmental Behavior of Pesticides*; Führ, F., Hance, R. J., Plimmer, J. R., Nelson, J. O. Eds.; ACS Symposium Series 699; American Chemical Society: Washington, DC, **1998**; pp 40-51.
9. Lee, J. K.; Führ, F.; Kyung , K. S. *Chemosphere*, **1994**, *29*, 747-758.
10. Casida, L. E. Jr. *Appl. Environ. Microbiol.* **1977**, *34*, 630-636.
11. Alef, K.; Kleiner, D. *Biol. Fertil. Soils.* **1989**, *8*, 349-355.
12. *Microbes in Action*; Seeley, H. W. Jr.; VanDemark, P. J.; Lee, J. J. Eds.; Fourth Ed.; W. H. Freeman and Company, New York, **1991**; pp 93-102.
13. Cho, K. H.; Sakong, J.; Kim, Y. K. *Agricultural Chemistry and Biotechnology* **1998**, *41*, 130-136.
14. Cox, L.; Koskinen, W. C.; Yen, P. Y. *Soil Sci. Soc. Am. J.* **1998**, *62*, 342-347.
15. Baruah, M.; Mishra, R. R. *Soil Biol. Biochem.* **1984**, *16*, 423-424.
16. Stevenson, I. L. *Can. J. Microbiol.* **1959**, *5*, 229-235.

Chapter 4

Metabolism of Ethaboxam in Soil

Jeong-Han Kim[1], Yong-Sang Lee[1], Young-Soo Keum[1], Min-Kyun Kim[1], Seung-Hun Kang[2], and Dal Soo Kim[2]

[1]School of Agricultural Biotechnology, Seoul National University, 103 Seodundong, Suwon 441–744, Korea
[2]Agrochemical Research Center, LG Chem Investment, 104–1 Moonjidong, Yusunggu, Taejon, Korea

Ethaboxam, N-[cyano(2-thienyl)methyl]-4-ethyl-2-(ethylamino)-1,3-thiazole-5-carboxamide, is a new systemic fungicide against *Oomycetes*. An aerobic metabolism study was carried out for 60 days with [^{14}C]ethaboxam applied at a concentration of 0.372 μg g^{-1} in a sandy loamy soil. The material balance ranged from 95 to 101% and showed a half-life of 17 days. The metabolites identified were N-[(Z)-amino(2-thienyl)methylidene]-4-ethyl-2-(ethylamino)-1,3-thiazole-5-carboxamide (AMETC), 4-ethyl -2-(ethylamino)-1,3-thiazole-5-carboxamide (ETTC), 4-ethyl-2-(ethylamino)-N-(2-thienylcarbonyl)-1,3-thiazole-5-carboxamide (ETC), and 4-ethyl-2-(ethylamino)-1,3-thiazole-5-carboxylic acid (ETCA). AMETC, ETTC, ETC and ETCA were recovered with a maximum level of 33, 10, 6 and 2% of the amount of applied radiocarbon, respectively, at 60 days after treatments. Evolved [^{14}C] carbon dioxide accounted for up to 18% and no volatile products were detected. Solvent non-extractable soil bound ^{14}C-residue reached 38% of the applied material at 60 days after treatment and radiocarbon in humin fraction was a major portion (20%) while humic acid and fulvic acid fraction retained 4 and 8%, respectively. Incubation of individual metabolite in soil with or without thiophene-2-carboxamide (TP) suggested that AMETC could be formed by the recombination of TP and ETC, which could be produced from the degradation of ETTC. All metabolites degraded in soil with half-lives ranging from 8 to 23 days.

Degradation of pesticides in soil is determined by many soil properties such as texture, pH, organic matter, moisture, and microorganisms (*1-5*). In addition to the fate of parent compound, metabolites are very important since they may persist in the environment or/and be equally or more toxic than the parent pesticides. Most of the fundamental information on the degradation of pesticides in soil and on the nature and extent of metabolite formation generally are obtained from laboratory soil metabolism studies to avoid the variation of many factors in field condition (*6, 7*).

Among the various laboratory systems, flow-through systems are the most popular approach to the pesticide metabolism studies in soil because the control of aeration and the trapping of carbon dioxide and volatile compounds are easily obtained to give the better mass balance (*8, 9*). Ethaboxam, N-[cyano(2-thienyl)methyl]-4-ethyl-2-(ethylamino)-1,3-thiazole-5-carboxamide is a new systemic fungicide recently developed by LG Chemical Ltd., Korea (Figure 1). This thiazole carboxamide was found to be very effective against *Oomycetes* which is responsible for late blight and downy mildew of various vegetables and fruit.

While *Oomycetes* is reported to have resistance to some current fungicides such as metalaxyl, this new fungicide showed high efficacy to those resistant *Oomycetes* (*10-13*) by inhibiting post-germination mycellial growth and the migration of nuclei from cyst to the growing mycellia. It showed good protective, curative, and persistent activity was observed in the growth chamber test (*14, 15*), and recent field studies revealed that ethaboxam wettable powder effectively controlled diseases caused by *Oomycetes* without any phytotoxicity (*16*). It has low mammalian toxicity and other adverse toxicological effects including mutagenecity and teratogenecity were reported (*15*).

This study was conducted to determine the fate of ethaboxam in soil under aerobic conditions. The degradation pattern of ethaboxam, structure and formation of the metabolites, and the distribution of soil bound residues were examined with a flow-through soil metabolism system.

Materials and Methods

Test Soil

Sandy loam soil was obtained from an upland field for the depths of 0-10cm. After air-drying at room temperature for 24 h, soil was sieved to remove any particles larger than 2 mm prior to characterization (Table I).

Table I. Physicochemical properties of the test soil

pH (1:10)	Organic matter (%)	CEC (meq/100 g Soil)	Field moisture at 1/3 bar (%)	Sand	Silt (%)	Clay	Texture
5.2	3.45	9.8	25.4	45.7	42.4	11.9	Sandy loam

Soil was adjusted to 75 % of field moisture content by adding the required amount of distilled water.

Radioisotope and Metabolites

[^{14}C]Ethaboxam labeled in thiazole ring and metabolites were kindly provided by LG chemical Ltd., Korea. The purity of [^{14}C]ethaboxam was examined by radioisotope-high performance liquid chromatography (RHPLC) just prior to the start of experiment (radiochemical purity, >99%; specific radioactivity, 26.2 mCi mmol^{-1}). The metabolites obtained were N-[(Z)-amino(2-thienyl)methylidene]-4-ethyl-2-(ethylamino)-1,3-thiazole-5-carboxamide (AMETC), 4-ethyl -2-(ethylamino)-1,3-thiazole-5-carboxamide (ETTC), 4-ethyl-2-(ethylamino)-N-(2-thienylcarbonyl)-1,3-thiazole-5-carboxamide (ETC), 4-ethyl-2-(ethylamino)-1,3-thiazole-5-carboxylic acid (ETCA) and thiophene-2-carboxamide (TP) (Figure 1).

Radioassay

Radioassay of all samples was done by triplicate. Radioactivity of liquid samples in Insta-gel scintillation cocktail (10 ml) was measured using a liquid scintillation counter (LSC) (Wallac model 1409, Turku, Finland). The solvent non-extractable residue (200-400 mg) was mixed with cellulose powder (100-200 mg), Combustaid (100-200 μl) and pelletized before combusted by a Packard model 307 oxidizer. The $^{14}CO_2$ produced was absorbed in Carbo-sorb E (10 ml) and mixed with Permafluor E^{++}(10 ml) for LSC counting.

Chromatography

RHPLC was performed by a Thermo Separation Product (TSP) model P2000 HPLC system with a C$_{18}$ column (μBondapak, 5 μm, 3.9 x 300 mm). A two step linear gradient was employed over 60 min with flow rate of 1.0 ml

68

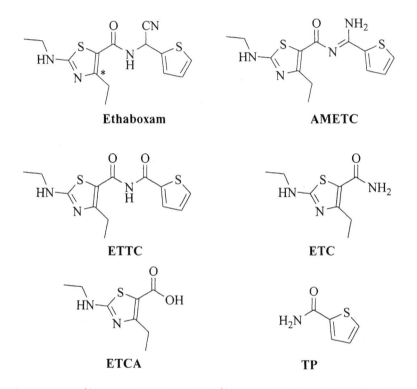

Ethaboxam

AMETC

ETTC

ETC

ETCA

TP

Figure 1. [^{14}C]ethaboxam (* Site of ^{14}C label) and related metabolites.

min^{-1}. The gradient was [10% acetonitrile (A), 90% of 0.2% TFA aqueous solution (B) at 0 min; 30% A, 70% B at 20 min; 70% A, 30% B at 40 min; 30% A, 70% B at 40 min]. UV detection was made at 310 nm and radioactivity monitoring (Berthold LB506 C-1 radioactivity monitor, 1.0 ml liquid cell) was performed using Scintillation cocktail (Flo Scint III, 3 ml min^{-1}). The identity of peaks was confirmed by co-chromatography with reference compounds.

Soil Incubation

Soil (50.0 g) in 100 ml metabolism flask was preincubated at 25 ($\pm$ 1)°C for three weeks in the dark with continuous air flow (5 ml min^{-1}) and [^{14}C]ethaboxam (1. 52 μCi in 300 μl acetonitrile) was applied dropwise to soil at a concentration of 0.372 μg g^{-1}. Soil samples were then incubated for up to 60 days at 25($\pm$ 1)°C with continuous air flow. Readjustment of soil moisture content (75% of field moisture content), and trapping of [^{14}C]carbon dioxide and volatile products were carried out once a week. Treated soil was extracted and analyzed at 0, 3, 7, 14, 30, 45, 60 days after treatment (DAT).

Extraction Efficiency

Soil samples (50.0 g) were treated with [^{14}C]ethaboxam (1.52 μCi) and incubated for 30 mins. The treated soils were extracted in a sequence with 100 ml of acetone, aqueous acetone (1/1, v/v) or acetone/water/36% HCl (35/35/2, v/v) by shaking for 1 h. The extract was decanted after centrifugation at 3500 rev min^{-1} for 10 min and the same procedure was repeated 3-4 times until the radioactivity of the extract reached the background level. The pooled extracts were adjusted to a known volume (400-500 ml) with acetone and aliquots of the extracts (1 ml) were radioassayed. The extract was then concentrated and the residue was dissolved in acetonitrile (5 ml) for RHPLC analysis.

Trapping of Volatiles and [^{14}C]Carbon Dioxide

Two polyurethane foam plugs and ethylene glycol (30 ml) were used as volatile traps, and [^{14}C]carbon dioxide was trapped by the two KOH solution traps (1 N, 30 ml). The foam plugs used to trap volatile products were extracted with methanol (30 ml). A sub-sample of the methanol extract (1 ml) and ethylene glycol (1 ml) from the trap were counted by LSC for radioactivity. To quantify evolved [^{14}C]carbon dioxide, aliquots (2 ml) of the two KOH traps were mixed with scintillation cocktail and stored overnight at 4°C before LSC counting. Levels of [^{14}C]carbon dioxide were confirmed by reacting the KOH solution (10 ml) of each samples with BaCl$_2$ solution (1 M, 10 ml) and the resulting supernatant was counted after the filtering of the [^{14}C]barium carbonate precipitate.

Extraction and Analysis of Soil Extract

After the trapping procedure was complete, soil samples were extracted with acetone (100 ml) and analyzed as in the procedure for extraction efficiency. In samples where more than 10% of the applied radioactivity remained in the soil after extraction, the sample was extracted with a mixture of acetone and water (1/1, v/v) and then a mixture of acetone, water and 36% aqueous HCl (35/35/2, v/v).

The remaining soil was air-dried, pulverized, and a sub-sample was combusted by a sample oxidizer.

Fractionation of Non-extractable Soil-bound Residues

Two grams of the non-extractable soil residues was extracted with NaOH solution (0.1 N, 5 ml) at room temperature under nitrogen atmosphere to minimize chemical change of humic substances due to auto-oxidation. The supernatant was decanted after the extract was centrifuged at 3,500 rev min^{-1} for 10 min. This procedure was repeated until the radioactivity of the extract reached the background level. Sub-samples of the precipitate were combusted to determine the radiocarbon content of the humin fraction. The extracts were combined, and concentrated and hydrochloric acid was added to it to adjust the pH to 1. This mixture was centrifuged, supernatant decanted and resulting precipitate (humic acid fraction) was washed with HCl solution (0.2 N, 5 ml). After centrifugation, supernatants (fulvic acid fraction) were combined and humic acid fraction was dissolved with NaOH solution (0.1 N, 5 ml). Aliquots (1 ml) of fulvic acid and humic acid fractions were radiocounted by LSC.

Formation and Degradation of Individual Metabolites

Each of the four metabolites (1 mg respectively) were incubated with TP (1 mg) for 60 days in soil (50g) and soil sample was analyzed at 0, 1, 3, 7, 14, 30, 45 and 60 days after treatment. Similar experiments were also performed with AMETC, ETC and ETTC, except for no addition of TP.

Results and Discussion

Material Balance

[^{14}C]ethaboxam was quantitatively recovered from soil with acetone and aqueous acetone without degradation. The material balance over the time course of the study was 95 - 101% of applied radiocarbon (Figure 2). Solvent extractable radiocarbon levels decreased gradually from 99.5% (0 DAT) to 40.6% (60 DAT) while non-extractable radiocarbon steadily increased to 33% of applied radiocarbon at 14 DAT and remained approximately constant until 60 DAT (37.6%) (Figure 3). The cumulative evolution of [^{14}C]carbon dioxide increased slowly and reached up to 18.2% of the applied radiocarbon, but no volatile products were detected suggesting that a significant amount of compounds have been mineralized into CO_2 or incorporated into humic materials without production of volatile metabolites.

The distribution of non-extractable radiocarbon in the humin, humic acid and fulvic acid fractions were examined (Figure 4) (*17-19*). Radiocarbon in the soil organic matter was present in order of humin (26%), fulvic acid (7.8%) and humic acid (3.5%) at 60 DAT.

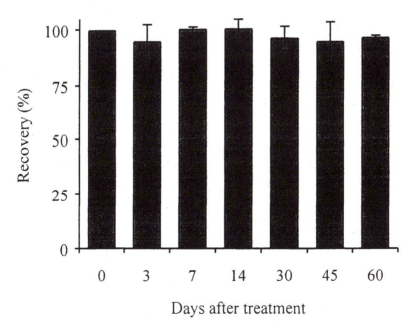

Figure 2. Material balance from the metabolism of [^{14}C]ethaboxam in soil.

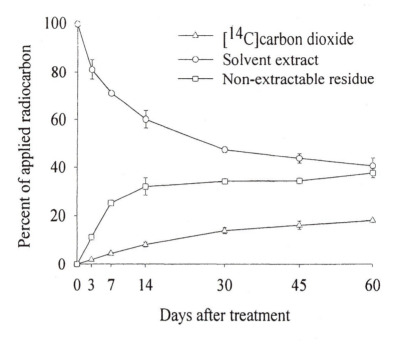

Figure 3. Radioactivity distribution in solvent extracts, [^{14}C]carbon dioxide and solvent non-extractable residue.

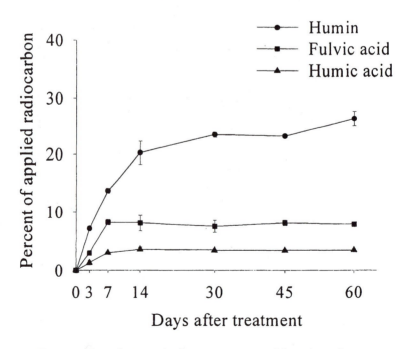

Figure 4. Distribution of solvent non-extractable radiocarbon.

Radioactivity increased until 7 days in fulvic and humic acid fractions and then kept constant. The level in humin fraction, however, increased continuously during the experiments.

Degradation of Ethaboxam

Ethaboxam was observed mainly in the acetone extract and exponentially declined to 5.2% of applied radiocarbon at 60 DAT ($Y = 93.3146e^{-0.2518X} +$ 6.8697; $r^2 = 0.9921$), giving the calculated half-life of 14 days (Figure 5-A). AMETC was detected as a major metabolite up to a maximum level of 32.9% of applied radiocarbon at 14 DAT, together with ETC, ETTC, ETCA as minor metabolites, of which the maximum concentration were 9.8% at 7 DAT, 5.8% at 3 DAT and 1.7% at 45 DAT, respectively (Figure 5-B).

Two or three small peaks were also observed, indicative of a diverse metabolic pathway was involved. The formation of ETTC through cyano group removal was expected as reported in other studies with pyrethroids (17), however, the formation of AMETC was not accounted for through a simple functional group transformation or degradation. Instead, it could be produced from the reaction of one of the metabolites, ETC, with TP which is a feasible product from the degradation of ETTC. Therefore, details of related mechanism were investigated through the incubation of individual metabolites and TP.

Formation and Degradation of Individual Metabolites

To investigate the formation mechanism of the major intermediate AMETC and the degradation pattern of individual metabolites they were incubated with or without TP in soil. AMETC was observed when ETTC was incubated without TP or with TP and when ETC was incubated with TP while no AMETC was found from the incubation of ETC without TP (Table II).

These results suggested that AMETC could be formed by the recombination of TP and ETC, which could be produced from the degradation of ETTC. The maximum level of AMETC (21.6%) was observed from the incubation of ETTC + TP (Figure 6).

When AMETC was incubated, to elucidate its fate, without or with TP, ETC was found as a major metabolite in the both cases together with small amount of ETCA. ETCA must be a terminal residue because it was found in all cases of the above incubations and no other compounds were found when it was incubated with TP. Degradation in soil was observed for all metabolites to give calculated half-lives of 23, 8, 14 and 10 days for AMETC, ETTC, ETC and ETCA, respectively (Figure 7).

On the basis of the results, a possible metabolic pathway for ethaboxam in soil under aerobic condition is proposed in Figure 8.

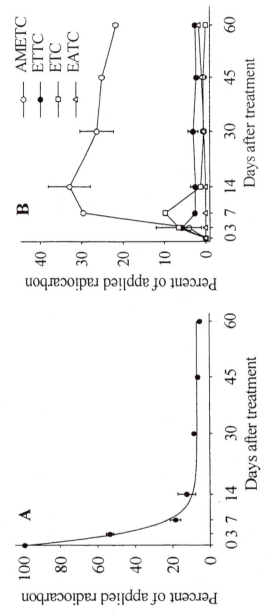

Figure 5. Degradation of [^{14}C]ethaboxam(A) and formation of metabolites(B).

Table II. Formation of Metabolites in soils when incubated for 14 days with or without thiophene-2-carboxamide (TP)

Treatment	(%) of treated compound			
	AMETC	ETTC	ETC	ETCA
AMETC	-	ND	6.7 ± 0.93	3.4 ± 1.4
AMETC + TP	-	ND	17.6 ± 6.0	1.2 ± 0.7
ETTC	3.9 ± 0.5	-	ND	0.9 ± 0.7
ETTC + TP	21.6 ± 1.9	-	6.1 ± 0.9	4.3 ± 1.0
ETC	ND	ND	-	4.3 ± 0.8
ETC + TP	6.4 ± 1.5	ND	-	2.8 ± 0.9
ETCA + TP	ND	ND	ND	-

ND: Not detected

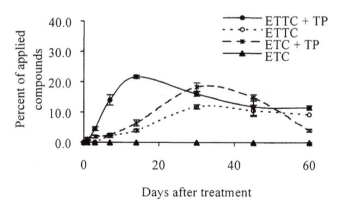

Figure 6. Formation of AMETC from ETTC or ETC, with or without TP in soil.

6
3
0
-3
Ln(mg/kg)
0
20
40
60
Days after treatment
AMETC
ETTC
ETC
ETCA

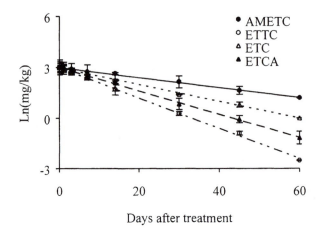

Figure 7. Degradation of individual metabolite in soil with TP.

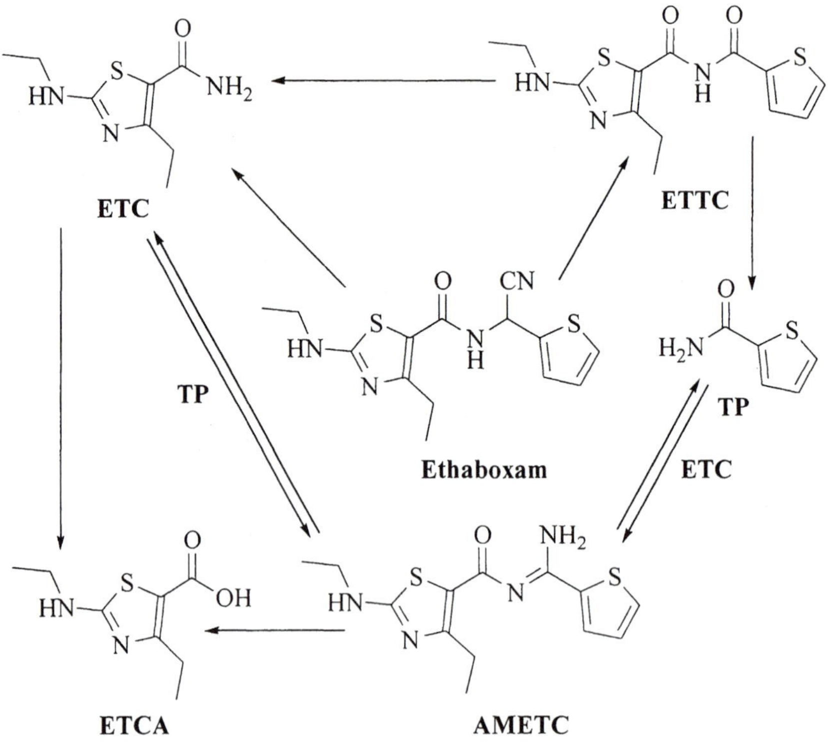

ETC
ETTC
TP
Ethaboxam
TP
ETC
ETCA
AMETC

Figure 8. Proposed metabolic pathway of [^{14}C]ethaboxam in soil under aerobic condition.

ACKNOWLEDGEMENT

This study was supported by SNU research fund and by Brain Korea21 project. Authors gratefully acknowledge Dr. Qing X. Li of the University of Hawaii at Manoa for his review of this manuscript.

REFERENCES

1. Athiel, P.; Mercadier, C., Vega, D.; Bastide, J.; Davet, P.; Brunel, B.; Cleyetmarcel, J.C. *Appl. Environ. Microbiol.* **1995**, *61*, 3216-3220.
2. Cink, J.H.; Coats, J. R. In *Pesticide Transformation Products, Fate and Significance*; Racke, K.D.; Leslie, A.R., Eds.; ACS Symposium Series 522, American Chemical Society: Washington D.C., **1993**, pp. 62-69.
3. Locke, M.A.; Harper, S.S. *Pestic. Sci.* **1991**, *31*, 221-237.
4. Simon, L.; Spiteller, M.; Haisch, A.; Wallnofer, P.R. *Soil Biol. Biochem.* **1992**, *24*, 769-773.
5. Talebi, K.; Walker, C.H. *Pestic. Sci.* **1993**, *39*, 65-59.
6. Nelson, H.P.; Termes, S.C. In *Agrochemical environmental fate*; Leng, M. L.; Leovery, E.M.K.; Zubkoff, P.L., Eds.; Lewis Publishers: New York, **1995**, pp. 51-69.
7. U.S. Environmental Protection Agency (US EPA). *Environmental fate*: Pesticide assessment guidelines; EPA-540/9-82-021, U.S. Government Printing Office: Washington D.C., **1982**, pp. 54-59.
8. Guth, J.A. In *Progress in Pesticide Biochemistry*, Vol 1; Hutson, D.H.; Roberts, T.R., Eds.; John Wiley & Sons: Chichester, **1981**, pp. 85-114.
9. Lee, P.W.; Steams, M.; Hemandez, H.; Powell, W.R.; Naidu, M.V. *J. Agric. Food Chem.* **1989**, *37*, 1169-1174.
10. Choi, G.J.; Kim, B.S.; Chung, Y.R.; Cho, K.Y. *Korean J.Plant Pathol.* **1992**, *8*, 34-40.
11. Ham J.H.; Hwang B.K., Kim, Y.J.; Kim, C.H. *Korean J. Plant Pathol.* **1991**, *7*, 212-220.
12. Hwang, B.K. *Plant Dis.* **1995**, *79*, 221-227.
13. Koh, Y.J.; Jung, H.Y.; Fry, W.E. *Korean J. Plant Pathol.* **1994**, *10*, 92-98.
14. Kim, D.S. *Biological activity of LGC-30473*; Working group on fungicide resistance, Korea, **1997**, pp. 7-64.
15. LG Chemical Ltd., Report, Korea, **1995**
16. Kim, D.S.; Park, H.C.; Chun, S.; Yu, S.H.; Choi, K.J.; Oh, J.H.; Shin, K.H.; Koh, Y.J.; Kim, B.S.; Hahm, Y.I.; Chung, B.K. *Korean J. Plant Pathol.* **1999**, *15*, 48-52.

17. Aizawa, H. *Metabolic maps of pesticides, Vol 1*; Academic press: New York, **1982**, pp. 179-207.
18. Khan, S.U. *J. Agric. Food Chem.* **1982**, *30*, 175-79.
19. Parsons, J.W. In *Humic substances and Their Role in the Environment;* Frimmel, F.H.; Christman, R.F., Eds.; John Wiley & Sons: Chichester, **1988**, pp. 3-14.

Chapter 5

Environmental Fate: Development of an Indoor Model Test for Runoff

Yoshiyuki Takahashi

Research Institute of Japan Plant Protection Association, Kessoku-cho 535, Ushiku, Ibaraki 300–1212, Japan

A new indoor test system with controlled rain conditions was developed to better understand field runoff. Containers ($0.7m^2$) packed with soil from the plow layer were used as test plots and placed under a rain-making machine. The test system was able to simulate a field runoff event. The runoff obtained by the test system approximates the runoff in the field. Some model tests using the indoor test system with sloped test plot, a mixture of three pesticides (chlorothalonil, diazinon and dimethoate), and a 30mm/hr rainfall were performed to detect the difference in runoff between cropped and non-cropped test plots. Furthermore, the test system was applied to prediction of water runoff in fields.

Background

Water runoff from croplands has been a major concern due to the soil erosion (1) that it causes and the potential hazard it poses by the transport of pesticides to public waters (2). Soil loss and erodibility have been studied not only in the field (*1*) but also in sloping lysimeters (*3*) and in soil beds packed in small pans (*4, 5*). In Japan, individual croplands are usually very small and scattered with different crops such as vegetables and fruit trees. These individual croplands generally do not have irrigation systems because there is sufficient rainfall throughout the year. Consequently it is difficult to estimate pesticide runoff from each individual field. It is for this reason that field runoff model tests have been performed (*6, 7*). The concept of model test is a single application of target pesticide to crops and collection of runoff water by heavy rainfall just after the pesticide application. It has been attempted to get the ratios of the runoff pesticides against applied amount of the pesticides.

Field model test for runoff with natural rainfall is unrepeatable, and it also has a seasonal limitation for heavy rainfall as well as geographical limitations (*7*). On the other hand, a field model test with artificial rainfall by irrigation systems needs a large amount of water (*6*). These characteristics indicate the field tests require a great deal of cost and labor. Therefore, obtaining data from the runoff events in the field has been difficult. Thus, an indoor test system has been developed with a rain-making machine (*8*). This paper presents the indoor runoff test system and the result of some model tests (*8*).

Outline of The Indoor Runoff Test System

Artificial rainfalls were produced by a rain-making machine (DIK-6000-S, Daiki Rika Kogyo Co., Ltd., Tokyo), which has an available rainfall area: $1.04m^2$ ($1.02m \times 1.02m$), rainfall intensity: 10-80mm/hr and raindrop size: 1.7-3.0mm diameter. Rainfall intensities of 12, 20, and 30mm/hr were used for runoff tests with 2.3mm diameter raindrops. Calibration of each rainfall was determined three times every 10min rainfall by the amount of precipitation into a $0.32m^2$ container.

Test plots were prepared by using containers ($0.7m^2$: 0.76m wide, 0.93m long and 0.2m in depth) which were packed with Andosols that covers more than 40% of upland field in Japan. Properties of the soil used are a clay loam (clay 17.4%, silt 29.9%, and sand 52.7%), organic carbon 3.42%, pH6.1/H_2O and 5.1/KCl, CEC 25.5mEq/100g, phosphate absorption coefficient 1,120, maximum water-holding capacity 81.8%. The containers were filled with soil and exposed to the artificial rainfall to remove soil cracks. After compaction via drying, additional soil was added to the containers. The soil surface was tilled thoroughly prior to testing. Each container was put on the carrier at a 5° angle. The test plot with container had a drainage system for percolating water and a collecting system for flowing surface water (Figure 1). Furthermore, a ceramic soil water meter (SPAS-pF33, DKK Corp., Tokyo) was put into the 15cm depth to measure the soil moisture (pF value) of each test plot. The sensors indicated a pF value of 2.9-3.0 in air and 1.4-1.6 in water. Thus, the pF value in each test plot was measured between these ranges.

The relationship between soil moistures of test plots just before rainfall (initial pF values) and ratios of runoff water calculated at 1L collection by 30mm/hr rainfall was investigated by the indoor test system. The results are shown in Figure 2. The test plots dried around pF3.0 need longer rainfall to collect 1L of runoff water. Thus, lower ratios of runoff water such as 2% were observed. On the other hand, the test plots with soil moisture around pF1.5 need less rainfall to collect 1L of runoff. Thus, higher ratios of runoff water such as over 10% were observed. When the initial pF value of each test plot was reduced to pF1.6 by the rainfall, water runoff was usually observed from the test plots. Furthermore, at least 30-60mm of accumulated rain precipitation was needed to obtain the pF1.6 for the test plots.

Verification of The Indoor Runoff Test System

For the verification of the indoor runoff test system, a field runoff occurred on Sep. 22, 1996, by a heavy rainfall of typhoon storm (7) was simulated by the indoor test system. As the model pesticides, a 70ml of pesticides mixture containing 400mg/L of chlorothalonil (TPN) and diazinon and 430mg/L of dimethoate was applied to the test plot using a hand sprayer at a rate of 100L/1000m^2. The rainfall by the typhoon (Figure 3A) was simulated by the rain-making machine (Figure 3B) just 2 days after the pesticides application. Water samples were collected as field runoff and analyzed by GC-MS (9).

The field size was 840m^2, which was 1,200 times bigger than the indoor test plot. The soil (Andosols) used for the indoor test plots was collected from this field surface. In the field runoff a 290L with 5hr rainfall, a 530L with 0.5hr rainfall and 600L with following 2hr rainfall in all 1,420L of runoff water were collected from the field (Figure 3A)(7). The results of simulated runoff by the indoor test are shown in Table I . When volumes of runoff water (L/m^2) from indoor test were compared with those of field test, both results were almost the same. Furthermore, the concentrations of three pesticides obtained by the indoor test were also very close to the results of the field test. These results indicate that the indoor runoff test system has the potential to simulate field runoff.

Runoff Model Tests by Indoor Test System

In the indoor runoff test system, a larger amount of runoff water collected from such a small test plot would bring higher ratio of pesticide runoff than that from a field. Thus, 1L of water sample per runoff event was collected from each test plot in the model tests. This is equivalent to 1.43L/m^2, which was almost corresponding to the result of typhoon storm (1.69L/m^2) (7).

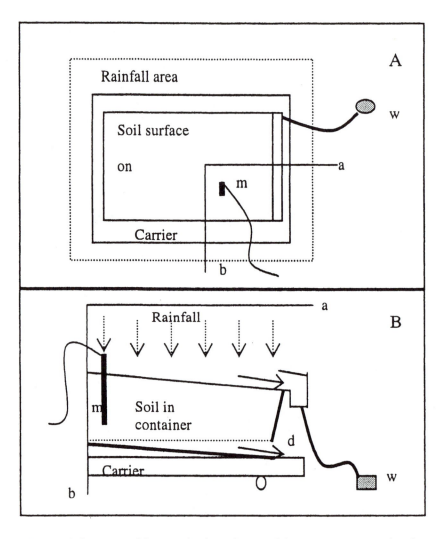

Figure 1. Structure of the test plot for indoor model test system. A: overhead view, B: cross sectional view (a-b), w: water sample collector, d: drainage, m: sensor for soil water meter. Reproduced with permission from reference 8. Copyright 2000 Pesticide Science Society of Japan.)

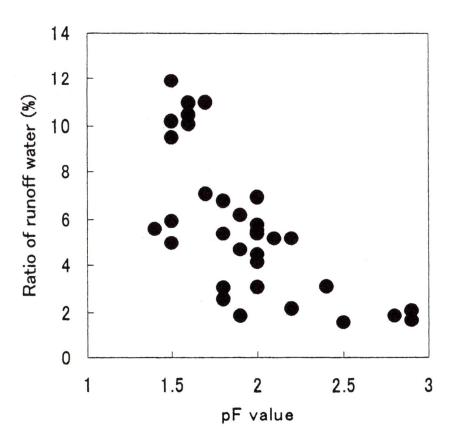

Figure 2. Relationship between soil moistures (pF values) and ratios of runoff water. (Reproduced with permission from reference 8. Copyright 2000 Pesticide Science Society of Japan.)

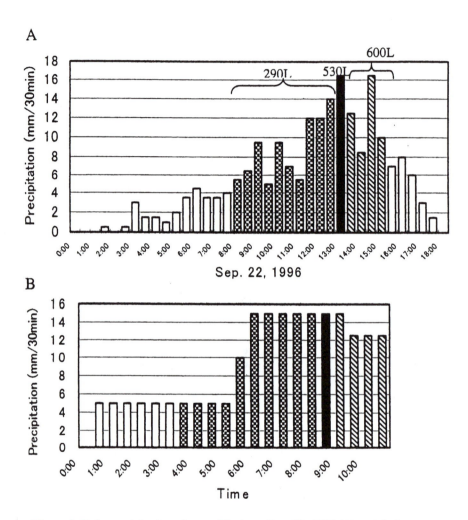

Figure 3. Rain precipitation of every 30min at Sep. 22, 1996 (A) and their simulated rainfalls by the rain-making machine (B). (A: Adapted with permission from reference 7. Copyright 2000 Pesticide Science Society of Japan.)

Table I. Simulation of Field Runoff

Runoff Test	Runoff Water (L/m²)	Concentration (mg/L)		
		TPN	Diazinon	Dimethoate
Field test (840m²)	0.345	0.005	0.001	0.002
	0.631	0.018	0.002	0.003
	0.714	0.008	0.001	0.002
Indoor test (0.7m²)	0.421	0.009	0.007	0.008
	0.653	0.010	0.005	0.002
	0.834	0.007	0.004	0.001

Runoff from Aged Test Plots

Individual test plots were aged for each period after pesticide application. Each 70ml of pesticide mixture diluted with tap water (TPN and diazinon: 400mg/L, dimethoate: 430mg/L) was applied to 5 test plots by using the hand sprayer at a rate of 100L/1000m². Each test plot placed under sunlight received 30mm/hr rainfall at 1 (No.1 test plot), 3 (No.2), 7 (No.3), 14 (No.4), and 21 days (No.5) after a single application of the pesticide mixture. Approximately 1L of runoff water was collected at each rainfall. Furthermore, a 210ml of the diluted pesticides mixture was applied to a test plot (No.6) for 3 times with 7 days intervals at a rate of 300L/1000m². After one hour from the last application, approx. 1L of runoff water was collected by 30mm/hr rainfall. All water samples were analyzed by GC-MS (9).

The concentration of pesticides in water samples from individual test plots (No.1 to No.5) decreases with time after pesticide application (See Figure 4). Furthermore, among the pesticides used, only TPN from test plots No.1 and No.6 showed a high ratio of runoff with suspended solids (SS). It seemed that TPN might be runoff with not only TPN absorbed SS but also flowable particles directly.

In many studies pesticide runoff was greatest when the rainfall occurred immediately and intensely after pesticide application (10, 11). Based on the results shown in Figure 4, the pesticide runoff from test plots of No.1 (single application) and No.6 (3 applications) were also greatest. Although the application dosage of the test plot No.6 was totally 9 times greater than other test plots, runoff TPN from the test plot No.6 was almost similar level of the No.1. On the other hand, runoff diazinon and dimethoate were approx. 2 times and 6 times greater than ones of the No.1. Thus, richer times and dosage of the pesticide application were not always increasing the concentration of runoff pesticides in water samples.

86

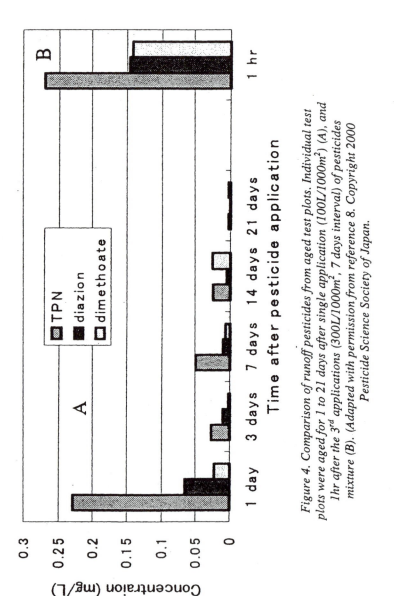

Figure 4. Comparison of runoff pesticides from aged test plots. Individual test plots were aged for 1 to 21 days after single application (100L/1000m^2) (A), and 1hr after the 3rd applications (300L/1000m^2, 7 days interval) of pesticides mixture (B). (Adapted with permission from reference 8. Copyright 2000 Pesticide Science Society of Japan.

Difference in Runoff Between Cropped and Non-cropped Test Plots

When cropped test plots are used, conditions such as deposit, uptake, washoff and degradation on foliages can be expected. Suitable plants for the test plot are of a short stature with many leaves to increase surface area for the spray application and reduce rebounding raindrops from leaves to out of the test plot. The French Marigold (*Tagetes patula* L.) was chosen as the crop. The indoor model tests for runoff were performed to detect the effect of crops on the test plots and for comparison to non-cropped plots.

A 300L/1000m^2 of diluted pesticides mixture (TPN and diazinon: 400mg/L, dimethoate: 430mg/L) was applied to 2 sets of non-cropped test plots (I-1 and I-2) and cropped test plots (II-1 and II-2), respectively. In each cropped test plot, 12 Marigolds (approx. 20cm high) were transplanted just 1 week before the pesticide application. Runoff tests with 30mm/hr rainfall were performed and 1L of water samples were collected from the same test plots after each period (1, 7 and 14 days after pesticide application), respectively. The water samples were analyzed by GC-MS.

The averaged ratios of runoff pesticides between the duplicated test plots are shown in Figure 5. Furthermore, the maximum ratio of pesticide runoff should be estimated from amount of runoff pesticides, which sequentially collected every day and integrated from a day after application to 14th days, as the worst case. The ratios of integrated runoff pesticide are presented in Table II. Based on the comparison of runoff between cropped and non-cropped test plots, both results were very similar in the ratios of integrated runoff pesticides. Furthermore, the ratios of integrated runoff pesticides were unexpectedly very low.

When crops existed in the test system, the soil moistures just before the rainfall and the ratios of runoff water were almost settled around pF2.0 and 5.3% with small changes (pF1.8-2.2 and 3.1-5.3%) between the duplicated test plots and between the runoff tests. However, the concentrations of pesticides in runoff water were maximally over 4 times different between the duplicated test plots of II-1 and II-2 (average 2.3 times). On the other hand, in the non-cropped test plots the soil moistures and ratios of runoff water changed more widely (pF1.4-2.9 and 1.7-5.6%) than ones of cropped test plots. However, the concentrations of pesticides in runoff water collected were maximally 1.8 times different between the duplicated test plots of I-1 and I-2 (average 1.3 times). The ratio of one test plot to the other one was detected between the duplicated test plots and indicated as differentiation. The runoff water ratio and runoff pesticide ratio such as TPN, diazinon and dimethoate were detected. Those results at 3 runoff events were averaged. The differentiations between the duplicated test plots are shown in Figure 6. Non-cropped test plots showed smaller differentiation than cropped test plots. Based on the results shown above, duplicated non-cropped test plots are recommended for the indoor runoff model test.

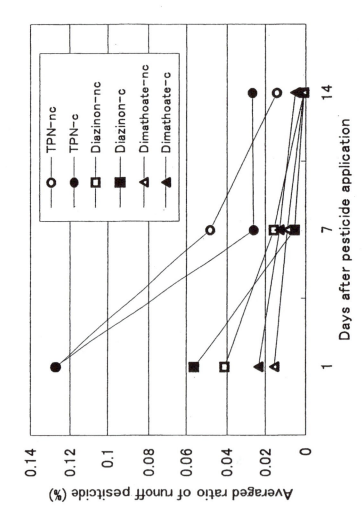

Figure 5. Ratio of runoff pesticides averaged duplicated test plots at the runoff events 1, 7 and 14 days after pesticide application with comparing between cropped (-c) and non-cropped (-nc) the test plots.

Table II. Ratios of Integrated Runoff Pesticides

Pesticide	Non-cropped	Cropped
TPN	0.75%	0.63%
Diazinon	0.25%	0.25%
Dimethoate	0.13%	0.16%

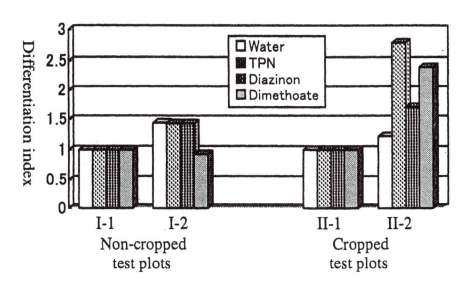

Figure 6. Differentiation between duplicated test plots. Ratios of runoff water and ratios of runoff pesticides were displayed.

Comparison of Runoff on Three Different Soils

Duplicated test plots packed with Andosols, Gray lowland soils which covers around 5% of upland field and Sand-dune Regosols which covers around 2% of upland field in Japan were used in the indoor runoff test. Soil properties of the Gray lowland soils are a clay loam (clay 24.5%, silt 26.7%, sand 48.7%), organic carbon 1.69%, pH 5.5/H_2O and 5.0/KCl, CEC 10mEq/100g, phosphate absorption coefficient (PAC) 889, maximum water holding capacity (MWHC) 62.0%. Soil properties of the Sand-dune Regosols are a sand (clay 0.7%, silt 11.9%, sand 87.4%), organic carbon 1.13%, pH 6.2/H_2O, CEC 9.1mEq, PAC 900, MWHC 49.5%.

When every 200ml of runoff water were collected continuously from the test plots by 12, 20, 30, and 40mm/hr of artificial rainfalls, collection times of every water samples (min/200ml/0.7m^2) were finally concentrated to the minimum collection times (runoff speed) as shown in Table III. Based on the results, field runoff may easily occur on a field of Gray lowland soils among the three soils.

Table III. Concentrated Collection Time of Runoff Water from Three Soils

Rainfall intensity (mm/hr)	Andosols	Gray lowland soils	Sand-dune Regosols
12	285.7	4.4	43.8
20	10.0 ·	3.2	12.0
30	4.0	2.0	4.6
40	2.0	1.3	2.2

NOTE: Each number indicates the minimum collection time for 200ml runoff water.

Those collection times (min/200ml/0.7m^2) were transformed equivalently to runoff water volume (L/1000m^2) per 10min rainfall at the four rainfall intensities. When the numbers were plotted in a figure, equations for water runoffs were obtained (Figure 7). The equation for Andosols was:

$$Ya = 0.5868X^2 + 14.822X - 252.02 \qquad (1)$$

For Gray lowland soils it was:

$$Yg = 1.3428X^2 - 13.052X + 613.44 \qquad (2)$$

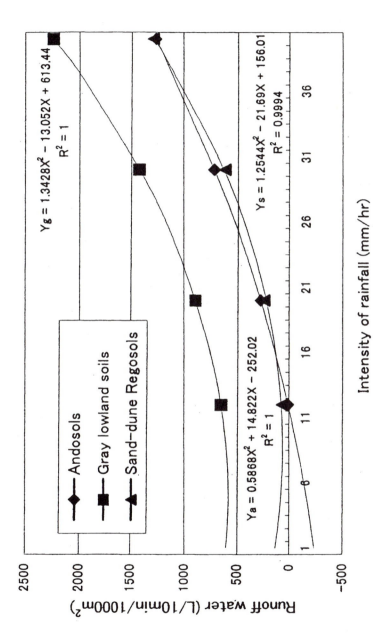

Figure 7. Runoff water from three tested soils at four rainfall intensities.

For Sand-dune Regosols it was:Ys= $1.2544X^2 - 21.69X + 156.01$. The "Ya", "Yg" and "Ys" indicate volume of runoff water (L/1000m^2) by 10min rainfall at the intensity of "X mm/hr".

Prediction of Field Runoff

On Field of Andosols

Occurrence of runoff on Andosols was deeply related with the soil moisture. Exactly, the pF value was required down to at least 1.6 for runoff (8). Thus, the equation (1), which is conditioned on saturated soil moisture, was unsuitable for prediction of field runoff on the Andosols. The concentrated collection time of 4min/200ml/0.7m^2 (Table III) is transformed equivalently to 0.72L/m^2/10min as the runoff speed. Further runoff speeds were determined by the other rainfall intensities. The results are shown in Figure 8. Based on the results, occurrence conditions for water runoff with each rainfall were arranged to the following. At 30mm/hr rainfall, water runoff started from 10min after the pF1.6 was obtained. The volumes of runoff water were 0.2L/m^2 at first 10min, 0.4L/m^2 at next 10min, and then finally 0.72L/m^2 at following 10min. At 20mm/hr, water runoff started from 30min after pF1.6 was obtained. The volumes of runoff water were 0.07L/m^2 at first 10min, 0.21L/m^2 at next 10min, and then finally 0.28L/m^2 at following 10min. At 12mm/hr, water runoff started from 240min after the pF1.6 was obtained. The volume of runoff water was maximally 0.01L/m^2/10min.

To predict runoff occurrence and amount of runoff water, rain precipitation data obtained from the field study in 1996 (7) were analyzed by using the occurrence conditions (Table IV). When the predicted amounts of runoff water and collected runoff were compared, the predicted amounts which were calculated by the maximum runoff speeds in the indoor tests were larger than the collected. However, the occurrence conditions obtained from the indoor test may be possible to predict the runoff events in field, though the conditions may be limited in Andosols.

On Gray Lowland Soils

In the indoor runoff test, the pF value of the soil was almost no change even though soil surface of the test plots were tilled thoroughly before testing. This may indicate that rainwater hardly percolates downward through the Gray lowland soils and occurrence of runoff on the soil may be independent of the soil moisture.When the "X" in the equation (2) was transformed to the rainfall

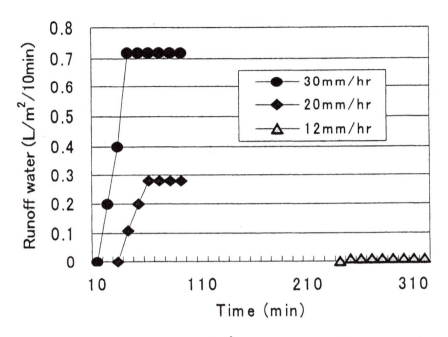

Figure 8. Amounts of runoff water (L/m²/10min) with 3 rainfall intensities (12, 20 and 30mm/hr). (Reproduced with permission from reference 8. Copyright 2000 Pesticide Science Society of Japan.)

Table IV. Prediction of Field Runoff on Andosols

Date of 1996	Size (m²)	Measured amount (L)	Predicted amount (L)
July 9	700	1.4	7
10	700	0.6	0
Sept. 22	840	1420	2318

NOTE: 7L= 0.01Lm²×700m², 2318L=(0.2+0.4+0.72×3)×840m².

intensity of 10min, it was reformed to: $Yg=48.298X^2 - 77.917X + 612.71$. Based on this equation, predicted amount of runoff water (PA) was calculated by the next equation: $PA = \Sigma(Yg)$. That is:

$$PA = \Sigma \, (48.298X^2 - 77.917X + 612.71) \tag{3}$$

The minimum point of "Yg" was obtained when "X" was 0.8 (mm/10min) in the equation. However, runoff events in a field (500m^2) of Gray lowland soils were mainly observed at over 1.5mm/10min rainfalls. Thus, the rainfall more than 1.5mm/10min in rainfall data of 10min interval precipitation from May to December in 1997 was put into the "X" of the equation (3). The results are shown in Table V.

Table V. Prediction of Field Runoff on Gray Lowland Soils

Rainfall in 1997	M: measured amount (L)	P: predicted amount (L)	Ratio of P/M (%)
May to July (Runoff: n=10)	11,178	11,671	104.4
Sep. to Dec. (Runoff: n=3)	13,776	16,520	119.9
Over 1.5mm/10min but no runoff (n=3)	0	3,767	-
Total	24,954	31,958	128.1

NOTE: Predicted amount (PA) was calculated by the equation (3). Field size was 500m^2.

The equation (2) was obtained from the results of the maximum speed and amount of runoff water on the indoor runoff tests. Thus, the result of predicted amount of runoff water by equation (3) should be larger than that of measured ones (Table V). Though further verifications are required, the equation (3) seemed to be useful to roughly estimate the amount of runoff water from fields of Gray lowland soils.

Conclusion

A small-scale runoff model test system with controlled rain conditions was developed and established as the indoor runoff model test. Though more tests with other kinds of soil are required, it seemed that the test system has the potential to simulate field runoff and to predict field runoff events. Furthermore, it would be useful for regulatory pesticide science, and may be useful for validating computer simulation models.

References

1. Crop Production Division, Agricultural Production Bureau, Ministry of Agriculture, Forestry and Fisheries: Chiryoku-hozen-taisaku shiryo (Countermeasure of maintenance of soil fertility), No. 55, pp1-14, 1979 (in Japanese)
2. Leonard, R.A. in Pesticides in the Soil Environment: Processes, Impacts, and Modeling; Cheng, H.H., Eds.; SSSA Book Series No.2, Soil Science Society America, Madison, 1990; pp 303-349.
3. Kamimura, H.; Shibuya, K. *Tech. Rep. Nat. Res. Inst. Agr. Eng.,* Series A, **1982**, *26*, 39-66. (in Japanese)
4. Truman, C.C.; Bradford, J.M. *Soil Sci.* **1990**, *150*, 787-798.
5. Bradford, J.M.; Huang, C. *Soil Technol.* **1993**, *6*, 145-156.
6. Takahashi, Y.; Odanaka, Y.; Wada, Y.; Minakawa, Y.; Fujita, T. *J. Pesticide Sci.* **1999**, *24*, 255-261.
7. Takahashi, Y.; Wada, Y.; Odanaka, Y.; Furuno, H. *J. Pesticide Sci.* **2000**, *25*, 140-143.
8. Takahashi, Y.; Wada, Y.; Odanaka, Y.; Kakuta, Y.; Fujita, T. *J. Pesticide Sci.* **2000**, *25*, 217-222.
9. Takeda, M.; Ito, Y.; Odanaka, Y.; Komatsu, K.; Maekawa, Y.; Matano, O. Nouyaku no Zanryubunseki Hou (Analytical Methods for Pesticide Residue), Chuokoki-shuppan, Tokyo, 1995; pp 141-143, pp 190-192, pp 226-227. (in Japanese)
10. Bowman, B.T.; Wall, G.T.; King, D.J. *Can. J. Soil Sci.* **1994**, *74*, 59-66.
11. Grynor, J.D.; MacTavisk, D.C.; Findlay, W.I. *Arch. Environ. Contam. Toxicol.* **1992**, *23*, 240-245.

Chapter 6

What Do We Know about the Fate of Pesticides in Tropical Ecosystems?

Kenneth D. Racke

Dow AgroSciences, Global Health, Environmental Science, and Regulatory,
9330 Zionsville Road, Building 308/2B, Indianapolis, IN 46268

Pesticide use is an important component of agricultural
and public health pest control in tropical areas. However,
the fate of pesticides in tropical ecosystems is not as well
understood as in temperate ecosystems. Although only a
few studies have directly compared pesticide fate in
tropical and temperate ecosystems, there is no evidence
that pesticides degrade more slowly under tropical
conditions. Laboratory studies under standardized
conditions reveal that pesticide degradation rate and
pathway are comparable between tropical and temperate
soils and waters. However, field investigations of tropical
pesticide fate indicate that dissipation often occurs more
rapidly than for pesticides used under comparable
temperate conditions. The most prominent mechanisms
for this acceleration in pesticide dissipation appear to be
related to the effect of tropical climates, and would
include increased volatility and enhanced chemical and
microbial degradation rates on an annualized basis.
Although the published literature contains a number of
reports and summaries of the behavior of pesticides in
tropical ecosystems, further experimental and modeling
research targeted at developing a more complete
understanding and better predictive capability of the
behavior of pesticides under tropical environmental
conditions should be encouraged.

Most investigations of pesticide environmental fate have been conducted in temperate ecosystems, predominately in North America, Europe, and Japan. Yet, approximately one-half the earth's population and roughly one-third of its land mass are found in the tropics. The countries of this zone, many of them developing, make substantial use of pesticides for control of agricultural and public health pests. In contrast, much less attention and funding has been directed toward understanding the fate of pesticides in tropical ecosystems. The primary objective of this review will be to explore answers to questions that may be posed relative to our knowledge of the fate of pesticides in tropical ecosystems. Although not designed as a comprehensive summary of available information, this review will use case studies and examples to highlight major trends and to offer a set of general conclusions and recommendations.

Why be Concerned About the Fate of Pesticides in Tropical Ecosystems?

First, and often of prime concern to pesticide users, are issues related to product efficacy. Will soil-applied herbicides, insecticides, and fungicides provide acceptable suppression or control of the target soil pests? Will foliarly-applied insecticides and fungicides protect the crop from the ravages of tropical plant diseases or destructive insect pests? There are some indications that, partially due to the aggressive nature of tropical pests and partially due to the harsh environmental conditions present in tropical environments, this may be a realistic concern. For example, applications of soil insecticides for termite control which typically provide 10-20 years of efficacy in temperate zones often provide only 2-5 years of control under tropical conditions (*1-3*).

A second concern regarding fate of pesticides in the tropics revolves around human exposure and health considerations. The distribution and persistence of pesticides within compartments of the tropical ecosystem are of major significance as relates to opportunities for exposure. For example, the persistence of pesticide residues on vegetable, fruit, and nut crop commodities bears directly on dietary intake considerations. In addition, the foliar behavior of pesticides is of interest from a worker reentry exposure standpoint. Finally, secondary exposure to pesticides may occur following offtarget movement in surface runoff or groundwater recharge to sources of drinking water.

A third concern regarding pesticides and their fate in tropical ecosystems involves possible impacts on ecosystem quality and diversity. The tropical regions are host to many of the richest biological communities on earth, and their vulnerability to human perturbations has not been thoroughly assessed. Included here would be both primary effects (e.g., acute toxicity) as well as secondary effects (e.g., food chain bioaccumulation). Although there are certainly more destructive and pervasive forces at work within the tropics (e.g.,

98

deforestation), understanding and minimizing the potential impacts of pesticides on tropical terrestrial wildlife and aquatic organisms should remain a priority.

It should be noted that entry of pesticides into tropical ecosystems can arise from either intentional introduction for pest management purposes or inadvertent entry resulting from improper waste handling or disposal operations. In fact, some of the countries facing the greatest challenges related to the disposition of obsolete pesticide stocks are located in the tropical region (*4-5*).

What Are the Tropics?

Tropical Environments

There are several ways in which tropical environments can be defined, with the geographic definition being the most commonly employed. The tropics can be geographically defined as that part of the world located between 23.5 degrees north and south of the equator, representing the landmasses between the Tropics of Cancer and Capricorn (*6*).

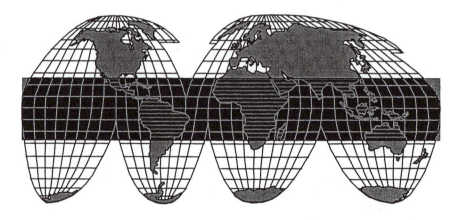

Figure 1. World map highlighting the tropical climate zone.

Tropical temperature regimes are warmer (year-round average) and exhibit much less variation in average temperatures from season-to-season as compared with temperate zones. The tropics have been characterized as that part of the earth where the mean monthly temperature variation is 5°C or less between the average of the three warmest and the three coldest months (*6*). Rather than view the tropics as a uniformly hot zone, the constancy rather than the absolute temperature of the tropics is the predominant distinguishing characteristic.

Given the relative uniformity of temperature, differentiation within the tropics is largely due to differences in the amount and distribution of precipitation. There are three fairly distinct tropical zones that can be delineated by moisture regime (6). In the low pressure belt around the equator, rainy climates prevail, creating a hot and humid zone which occupies approximately one-quarter of the tropical landmass. This udic moisture regime is characterized by large amounts of rainfall nearly evenly distributed throughout the year (7). Moving away from the equator, there is a tendency for the amount of rainfall to decrease, and for it to be unevenly distributed with one or two distinct dry periods per year. The ustic or seasonal moisture regime represents roughly one-half the landmass of the tropics, and includes those areas which experience classic monsoon climates. The aridic moisture regime, which prevails in approximately one-quarter of the tropical landmass, is characterized by either relatively short rainy seasons (dry climates) or sporadic precipitation (deserts) (7).

Another characteristic that differentiates tropical and temperate zones is the level of solar radiation. The mean daily incident solar radiation reaching tropical areas is roughly twice that of temperate areas. The increased solar radiation is due to several factors related to the orientation of the earth and sun, including the passage of the sun's rays through a thinner atmosphere in the tropics due to the more perpendicular angle. For example, nearly 60% of the incident sunlight passes through the atmosphere at the equator compared with 33-46% in the mid-latitudes. More ultraviolet rays also reach the earth's surface in the tropics (6).

Given some of these generalizations about tropical regions, it should be noted that tropical areas are diverse and do not easily fit generalizations due to other variables. Elevation, for example is one factor that can have as much affect as latitude on tropical climates. In spite of the geographic definition of the tropics, there also is no true dividing line between tropical and temperate zones. In fact, areas intermediate in character between tropical and temperate areas are often referred to as "subtropical".

Tropical Soils

The soils of the tropics, as those in the temperate regions, are highly diverse and strongly site dependent. Recent studies have demonstrated that tropical soils exhibit as broad a range of properties as soils of the temperate region (6). Tropical soils are not uniquely different than soils of the temperate zones, and rather than describe a specific set of soil properties, the term "tropical soil" refers to any soil that occurs in the tropics (8-9). Many, but not all, tropical soils are very old. Thus, many tropical soils have evolved to contain variable charge clay mineral systems that confer some distinct physical and chemical properties (8). Several generalizations that have been drawn about the character of soils which occur in the tropics are listed below (6):

- The kinds and properties of clay minerals are much more varied in the tropics than in glaciated temperate areas.

- Many tropical soils exhibit significant anion exchange capacity.

- Organic matter contents in the tropics are similar to those of the temperate region.

- Although the annual addition of organic carbon to the soil is five times greater in tropical udic environments than temperate udic environments, the rate of organic decomposition is also five times greater in the tropics.

- In ustic environments, lack of soil moisture during the dry season decreases organic carbon decomposition just as low temperatures do in temperate regions.

- The vast majority of the soils of the humid tropics are acidic.

- The vast majority of the cultivated soils of the humid tropics are not acidic.

The most widely used classification system for soils of the tropics is the U.S. Soil Taxonomy (10). Soil types (and their relative distribution) present the tropics include oxisols (22.5%), aridisols (18.4%), alfisols (16.2%), ultisols (11.2%), inceptisols (8.3%), and entisols (8.2%) (6,11-12).

Tropical Ecosystems

Due to the variability in environmental conditions of the tropical regions, especially that associated with moisture regime (udic, ustic, aridic), tropical ecosystems are quite varied and few generalizations can be offered. The natural vegetation of the tropics is closely correlated with climate, and it can be grouped into five general categories (6). These categories (and their relative distribution) include savannas and grasslands (43%), broad-leaved, evergreen rainforests (30%), semi-deciduous and deciduous forests (15%), desert shrubs and scattered grasses (7%), and no vegetation (5%) (6). Although tropical rainforests comprise only a portion of the tropical region, as opposed to conventional wisdom, their biotic diversity is truly remarkable. For example, almost 2500 tree species are found in the Malaysian and Amazon jungles as compared with about 12 species in temperate forests (6). The terrestrial and aquatic wildlife of the tropics are also quite diverse, as might be expected from the range of climatological and vegetative habitats available.

How Are Pesticides Used in the Tropics?

Tropical Agricultural

Although certain agricultural crops grown both in tropical and temperate zones (e.g., maize, wheat, potatoes), some are largely found in the tropics. Examples of such tropical crops would include cassava, yams, millet, bananas/plantain, sugarcane, coffee, pineapple, and oil palm. Most crop production in the tropics deals with one of two major farming systems (7,13). The first, adapted for the wet, equatorial tropics, involves the root and tuber farming system. The main source of food energy is from vegetatively propagated roots and tubers such as sweet potatoes, yams, and cassava, or fruits such as bananas and plantains. The second system, adapted for the seasonally dry tropics, involves cereal farming. Here the main sources of food are cereal crops such as sorghum, millet, and maize. Rice predominates in the monsoon regions of Southeast Asia. In addition to field crops, cultivation of vegetables, fruits (e.g., mango, pineapple), and fiber crops (cotton, jute) are also important in tropical areas.

Tropical agriculture, although traditionally less dependent on the use of pesticides than temperate agriculture, is increasingly relying upon modern agricultural approaches, including the use of pesticides. Several countries with substantial tropical area which rank among the leading world agrochemical markets (14). These would include Brazil (4th), Australia (11th), India (12th), PRC (13th), Mexico (16th), Colombia (17th), and Thailand (19th). It is difficult to estimate accurately what percentage of world pesticide use occurs in tropical areas, but it would probably represent on the order of 10-20% (15-17). Many tropical countries employ primarily insecticides as compared with other types of pesticide products (e.g., India). This is in contrast to most major markets in North America and Europe which are heavily focused on herbicides (17). The few tropical countries which do rely more heavily on herbicides often do so because export crops are heavily treated (e.g., Brazilian soybeans). It should be noted that several tropical countries, most notably India and Brazil, boast a significant local production capacity for pesticide products (14).

Public Health

In addition to agricultural uses, there are also important non-agricultural uses for pesticides in the tropics. Many of these uses are related to public health protection, and are targeted at control of disease-carrying insect vectors. The resurgence of such insect-transmitted diseases as malaria, leishmaniasis, and yellow fever, and the emergence of newly important diseases, has generated a renewed emphasis on pesticide use as a critical component in vector control programs (*18*). These vector control programs, which may involve pesticide application to surface water to control pest larvae or terrestrial environments to control adult pests, often rely upon older products such as the chlorinated hydrocarbon insecticides (e.g., DDT) that are no longer employed for agricultural pest control (*19*).

The importance of pesticides for control of wood-destroying insects in tropical areas is also well known. For example, in some tropical areas significant quantities of soil applied insecticides are employed to create termiticidal barriers around susceptible structures (*12*). Creation of these types of barriers may involve the application of several kg of active ingredient to the soil environment, and in tropical regions these applications may involve use of older, chlorinated hydrocarbon insecticides (e.g., chlordane). In contract, newer products (e.g., chlorpyrifos, bifenthrin) have replaced these chlorinated hydrocarbons in temperate zone countries.

Table I. Pesticide Market Comparisons for Leading Temperate Zone and Tropical Zone Countries

	U.S.	Japan	France	Brazil	India	Thailand
Market Division						
Insecticides	17%	34%	15%	21%	74%	34%
Herbicides	67%	32%	40%	56%	14%	52%
Fungicides	7%	31%	41%	16%	11%	12%
Crop Market Split						
Rice		37%		4%	31%	19%
Maize	27%		10%	9%		
Cereals	7%	3%	47%	3%	6%	
Soybeans	22%			34%		
Fruits/Vegetables	14%	48%	30%	21%	15%	42%
Sugarcane				11%		
Cotton	11%			4%	41%	
Coffee				7%		
Plantation Crops						20%
Other	19%	12%	13%	7%	7%	19%

SOURCE: Adapted from Reference 14.

Regulation of Pesticides in Tropical Areas

The tropical zone includes portions of more than 70 countries, which may range from highly developed to developing in nature. Some countries of the tropics have sophisticated and data-intensive pesticide registration processes. These processes typically involve establishment of registration guidelines and study requirements, submission of regulatory testing data by agrochemical manufacturers, and data review and risk management decisions by one or more independent government agencies. In addition to evaluation of new pesticide products, these countries also have tended to establish regulatory compliance monitoring and review processes for older pesticide products (i.e., reregistration). Examples of such countries would include Australia, the United States, Brazil, and Taiwan.

Other countries within this region have more recently established regulatory processes for the evaluation, registration, and periodic reevaluation of pesticide products. Many of these countries have been steadily moving toward regulatory processes which are allowing decisions to be made which are more sensitive to local needs, agricultural practices, and environmental conditions than ever before. Examples of such tropical countries would include Mexico, Colombia, India, and the Philippines. In some instances, advances in pesticide registration processes have involved partnerships with larger countries with longer regulatory experience, such as has occurred among Mexico and the U.S. and Canada via implementation of the North American Free Trade Agreement.

Finally, some countries have few available resources for comprehensive and/or formal pesticide evaluation and approval processes. In these countries, certain aspects of pesticide use may be focused on to the exclusion of others. For example, much attention may be placed on the country of origin for technical product, but little attention may be paid to ensuring availability of locally relevant residue or environmental fate data. In other cases, approval may be dependent upon demonstration that registration for a particular product exists in another country with a more advanced pesticide registration system (e.g., EU, U.S., OECD member).

Regarding the most common pesticides in use in the tropics, many of these are the same as those employed in temperate areas. One noteworthy difference, however, involves the use in tropical countries of older pesticide products which, in some cases, have fallen from common use in temperate zone countries due to restriction or replacement. One factor leading to this difference relates to economics. Newer and proprietary pesticide products may have a higher cost per use as compared to existing and often generically produced pesticides, and the agricultural economies of tropical zone countries may not merit this higher cost. An additional factor for this difference involves regulatory restrictions which may originate and be implemented first in temperate zone countries. For example, a number of pesticide products which still enjoy widespread use in tropical areas, such as DDT, monocrotophos, and parathion, have been severely restricted or banned from many other areas of the temperate zone (*14,19*).

How Much Information is Being Generated on the Fate of Pesticides in Tropical Ecosystems?

After reviewing the literature, it is clear that the bulk of pesticide fate research has been focused on the temperate zone. For example, a recent literature search of select citation indices (Chemical Abstracts, Biosis, SciSearch) for publications on the fate of commonly used members of several insecticide classes (1985-1996) revealed a disproportionate share of information from temperate soil research (*20*). Out of 445 citations a total of 341 were focused on temperate zone conditions (77%), and of those reports focused on the tropical zone the majority (80 reports or 18% of the total) originated in one country, India. This leaves only about 5% of the total which were focused on either the fate of pesticides in the tropical zone of other countries.

The reduced research focus on pesticides in tropical areas is probably due to several reasons, including greater requirements for regulatory testing and more plentiful academic research funds in temperate zone countries. Fortunately, there have been several research groups in tropical and subtropical areas that have been quite active over the years in generating significant pesticide fate information, both in terrestrial and flooded (i.e., rice paddy) agricultural environments. This would include contributions from the Central Rice Research Institute in India, Indian Agricultural Research Institute, Asian Vegetable Research and Development Center in Taiwan, International Rice Research Institute in the Philippines), Centro de Radioisotopos, Instituto Biologico in Brazil, the University of Hawaii, and the University of Costa Rica.

One opportunity highlighted by the preponderance of temperate zone research data on pesticide ecosystem fate relates to the applicability of such research toward an understanding of pesticide fate in tropical ecosystems. This topic is discussed in the following section.

How Does Pesticide Transformation Under Tropical Conditions Compare With That For Temperate Conditions?

With the bulk of pesticide ecosystem fate research data generated in the temperate region, whether of academic or regulatory origin, the applicability of these results for tropical ecosystems is of concern. Of particular interest are the potential similarities and differences of pesticide behavior in tropical versus temperate ecosystems. Although anecdotal comparisons or conclusions have been made, the best evidence of the comparability of temperate and tropical pesticide data would arise from either mechanistic laboratory investigations comparing fate of pesticides under simulated conditions in temperate and/or tropical matrices (e.g., soil, water) or from observational field dissipation studies in which a common study protocol was implemented both in tropical and temperate areas. Fortunately, a few such investigations have been completed (*21-24*), and results of these studies and other relevant investigations have been summarized in the following sections.

Terrestrial Ecosystems

Fenamiphos

The organophosphate fenamiphos is a soil nematicide used worldwide on a great number of agricultural crops. The degradation of ^{14}C-fenamiphos was examined in 16 soils (Table II) from both temperate and tropical/subtropical regions (Brazil, Costa Rica, USA-Florida, Japan, Thailand, Philippines) (*21*). Samples of each soil were treated with the nematicide at 7.7 ppm and incubated under aerobic conditions for 15, 50, and 90 days. Sets of temperate soils were incubated at both 16 and 22°C, whereas tropical/subtropical soils were incubated at 22 and 28°C. Under identical temperature conditions there was no discernible difference in quantities of fenamiphos TTR (total toxic residues = fenamiphos + f. sulfoxide + f. sulfone) or degradates remaining in the two soil groupings at similar time points. For soils maintained at 22°C, TTR remaining after 90 days in the 9 temperate soils was 28 ± 20% (SD) (range of 2-67%) and TTR remaining in the 7 tropical/subtropical soils was 32 ± 15% (range of 12-51%). Quantities of radiocarbon mineralized after 90 days were also similar for the temperate (20 ± 12%) and tropical/subtropical (13 ± 9%) soils. However, fenamiphos TTR remaining in temperate soils held at 16°C and tropical/subtropical soils held at 28°C were significantly different. Twice as much fenamiphos TTR remained in the temperate (45%) versus the tropical/subtropical soils (22%), and mineralization was also significantly reduced under the cooler conditions. Thus temperature seemed to have a more significant impact on degradation kinetics than did soil origin (i.e., tropical vs. temperate). In soils from both regions the pathway of fenamiphos degradation and the metabolites identified were the same, and the authors concluded that the main degradation pathway of a pesticide can be deduced with sufficient accuracy from examination of very few soils.

Atrazine

The triazine herbicide atrazine is commonly employed in many countries for control of broadleaf and grass weeds in maize, sugarcane, and other field crops. A series of experiments on atrazine degradation in tropical (Thailand) and temperate (Japan) soils was conducted by Korpraditskul et al. (*22,25*). In two separate experiments, samples of natural (nonsterile) soils were treated with

Table II. Degradation of ^{14}C-Fenamiphos in Soil
Under Laboratory Conditions.

Soil Origin	TTR^a % of applied at 90 days		Evolved ^{14}CO$_2$ % of applied at 90 days	
	22°C	28°C	22°C	28°C
Temperate				
Canada	34		17	
Sweden	36		16	
Germany-Bavaria	8		33	
Germany-R. Pfalz	29		13	
Netherlands	14		39	
France	2		32	
USA-Indiana	16		14	
USA-Nebraska	43		13	
Japan-Toyoda	67		1	
Mean	*28*		*20*	
Tropical/Subtropical				
USA-Florida	43	33	5	9
Costa Rica	30	24	12	16
Brazil-P. Fundo	18	8	21	37
Brazil-Parana	12	6	18	34
Thailand	51	40	4	10
Philippines	25	14	24	40
Japan-Tsurug	47	30	4	10
Mean	*32*	*22*	*13*	*22*

SOURCE: Data from Reference 21.

^{a}TTR = Total Toxic Residue (fenamiphos + f. sulfoxide + f. sulfone)

atrazine at 3 ppm and aerobically incubated for up to 90 days at 30°C (Figure 2). Korpraditskul et al. (*22*) first examined atrazine persistence in 5 Thai soils, and observed half-lives of 6-150 days. The 2 soils with lowest pH exhibited the most rapid degradation. Further work of Korpraditskul et al. (*25*) involved five Japanese soils (temperate) and 2 Thai soils (tropical) and was designed to determine the effect of soil properties on atrazine degradation. Half-lives in the soils ranged from 20-150 days, and were highly correlated with soil pH (r = 0.79). A comparison of results from both studies reveals no clear differentiation between the tropical and temperate soils based on observed rate of atrazine degradation. An additional environmental factor examined by Korpraditskul et al. (*22*) was temperature. Samples of 2 of the tropical soils were, in addition to 30°C, also incubated at 15, 25, 37, and 45°C. Atrazine remaining after 90 days was less in soils incubated at higher temperatures, and indicated that temperature is an important variable in the observed rate of degradation of this compound.

Further comparison of sterile and nonsterile samples of temperate and tropical soils revealed similar atrazine degradation rates, thus highlighting the importance of abiotic degradative mechanisms (*25*).

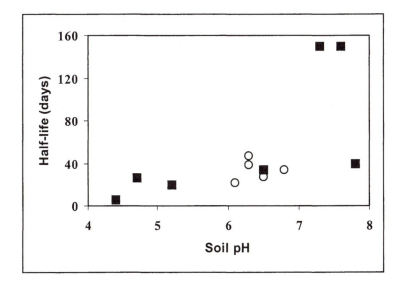

Figure 2. Degradation of atrazine in temperate (○) and tropical (■) soils under laboratory conditions (data from References 22 and 25).

Simazine

The comparative degradation of simazine as affected by temperature and moisture was investigated in a series of 16 temperate and tropical/subtropical soils (Taiwan, Philippines) in a factorial laboratory experiment (*23*). Soil samples were fortified with 4 ppm simazine and incubated for up to 140 days at 20-90% field moisture capacity and 5-45°C. Simazine half-life ranged from 11-476 days, and was significantly correlated with soil organic carbon content, clay content, sand content, and pH. At 25-30°C and 90% FC half-lives in temperate and tropical/subtropical soils were 17-76 (mean = 40 ± 19) days and 25-67 (mean = 41 ± 19) days, respectively.

In a companion effort, simazine persistence under field conditions at 21 sites in 11 countries was also examined. Soil surface applications of simazine at 2-4 kg ai/ha were made to the plots, and soil cores were taken for analysis to a depth of 10 cm. Field dissipation of simazine was most rapid under tropical and subtropical conditions, although a few temperate sites also displayed rapid dissipation (Table III). This was in contrast to the laboratory data, which had indicated no propensity for the tropical soils to induce greater rates of simazine degradation than the temperate soils.

The authors of this study attempted to employ Walker and Barnes' mathematical soil degradation model (*26*) to determine if prediction of the observed field behavior of simazine could be predicted based upon degradation rates from the laboratory and temperature and rainfall patterns recorded at the field sites. Occasionally there was close agreement between predicted and observed simazine residues, but in general the model tended to underestimate the rate of loss under field conditions. A further comparison between predicted and observed residues at all sites revealed an overall average standard deviation of ± 42.5%. These results highlight the difficulties of extrapolating laboratory data to explain field behavior. However, the meager fit of the best multiple regression analysis of laboratory half-life vs. soil properties (%OC + %clay + pH; $r^2 = 0.64$) also indicates difficulties that may be inherent in predicting even laboratory behavior under standardized conditions. This finding has significant implications for attempts to extrapolate results from temperate soils to those of tropical soils.

DDT

DDT still finds significant use in many tropical countries, especially for control of disease vectors. The degradation and persistence of ^{14}C-DDT under field conditions was the subject of a series of collaborative efforts in 14 countries sponsored by the International Atomic Energy Agency (*24,27*). Sites included primarily tropical and subtropical areas. The first set of experiments were conducted during the period 1982-1987, and the second set of experiments

Table III. Dissipation of Simazine in Soil Under Laboratory
and Field Conditions.

Location	Lab half-life (days)[a]	Field DT50 (days)[b]
Temperate		
Warwick, England	29	70-80
Saskatchewan, Canada	78	30-40
Firenze, Italy	31	20-30
Uppsala, Sweden	76	30-40
Braunschweig, Germany	42	40-50
Alberta, Canada	59	50-60
Oxford, England	26	90-100
Ontario, Canada	30	30-40
Ontario, Canada	33	20-30
Wageningen, Holland	27	70-80
Maarn, Holland	17	10-20
British Columbia, Canada	28	20-30
Harpenden, England	ND	>120
Maidstone, England	ND	40-50
Horotiu, New Zealand	ND	<10
Hamilton, New Zealand	ND	10-20
Copenhagen, Denmark	ND	70-80
Tropical/Subtropical		
Taipei, Taiwan	25	10-20
Taichung, Taiwan	31	10-20
Laguna, Philippines	67	<10
Bogor, Indonesia	ND	10-20

Source: Data from Reference 23
[a] 90% field capacity and 25-30°C
[b] Estimated from data

during 1989-1993. The studies were conducted under a standardized protocol for field and analytical aspects. At the majority of tropical sites dissipation of DDT occurred at a substantial rate (Table IV). After 12 months the quantity of DDT remaining in soil at tropical sites ranged from 5% of applied in Tanzania to 15% of applied in Indonesia. Likewise, DT_{50} values for total DDT residues (DDT + metabolites) ranged from 22 days in Sudan to 365 days in China. One exception was provided by an extremely acidic (pH 4.5) tropical Brazilian soil which yielded a DT_{50} of >672 days. Comparable DT_{50} values for DDT in

Table IV. Field Dissipation of ^{14}C-p,p'-DDT from Surficial Soil.

Site	Soil pH	Half-Life[a] (days)	DT50[a] (days)	Citation
Kenya-highland	6.3	65	23	35
Kenya	-	78-90	54-62	27
Tanzania-lowland	6.2	174	23	36
Tanzania-highland	6.9	335	170	36
Sudan	-	35	22	27
Egypt	-	224	130	27
India-lowland	7.9	319-343	120-125	37
India-highland	8.0	136	60	37
India	-	234	60-120	27
Pakistan-lowland	7.9	144	90	38
Pakistan-highland	8.1	313	240-300	38
Pakistan	-	112-120	75-90	27
Malaysia	-	105	30	27
Indonesia-lowland	5.7	236	175	39
Indonesia-highland	5.3	159	63	39
Philippines	6.1	210-261	82-100	40
China-highland-subtropical	6.1	525	>300	41
China-highland-tropical	7.2	-	204	41
Panama-highland	6.1	-	135	42
Panama-lowland	5.5	-	365	42
Brazil-tropical	4.5	1435	>672	43
Brazil-subtropical	4.8	>1400	>672	
Brazil	4.5	>800	320	27
USA-Hawaii-lowland	5.4	-	175	44
USA-Florida	<5.0	>678	340	27

[a] Dissipation of total ^{14}C residue: Extractable DDT, DDE, DDD, and soil-bound residue

temperate regions of 837-6087 days have been previously reported (*28-31*). Woodwell et al. (*32*) concluded that the mean lifetime of DDT in temperate U.S. soils was about 5.3 years. A major conclusion of the present study was that DDT dissipates much more rapidly in soil under tropical conditions than under temperate conditions. The major mechanisms of dissipation under tropical conditions included volatilization, biological and chemical degradation, and to a lesser extent binding to the soil matrix. However, within specific tropical countries evidence was generated to indicate that there could be large differences in degradation rates of DDT in soil due to the different climates and soil types. For example, DDT dissipated more rapidly from lowland (vs. highland) soils in Tanzania and Pakistan, but more rapidly from highland (vs. lowland) soils in India and Indonesia.

The primary metabolite of DDT detected in tropical soils was DDE, and its dissipation was also examined in 8 different countries. With the exception of Brazil (highly acidic soil), studies reported overall DDE half-lives of 151-271 days. Again this is much shorter than observed DDE half-lives of >20 years from temperate areas (*33-34*).

Sediment/Water Ecosystems

Given the importance of rice as a tropical food crop, it is not surprising that considerable attention should have been devoted to investigations of the fate of pesticides in flooded rice paddy soil under tropical conditions (*45-46*). A particularly useful summary of the fate and ecotoxicology of pesticides in the tropical paddy field system was prepared by Abdullah et al. (*47*). As with terrestrial ecosystems, however, few direct comparisons have been made between pesticide fate in sediment/water tropical and temperate soils. Flooded, paddy conditions usually result in a reduced soil layer, which significantly impacts the soil microbial community. The main biochemical processes in flooded soils can be regarded as a series of oxidation-reduction reactions mediated by different types of bacteria (*46*). A common result of the presence of these reductive conditions is that more rapid degradation of chlorinated hydrocarbon and nitro-containing pesticides is observed than under aerobic conditions. In addition to degradation in soil, the presence of a water layer increases the opportunity for hydrolytic and photolytic pesticide transformations.

Lindane

Lindane is still one of the most widely used chlorinated hydrocarbon insecticides in tropical areas, and due to extensive use in rice culture its fate in flooded soils has been fairly well studied. Several researchers have examined the fate of lindane in flooded soil under laboratory conditions (Table V). Absolute comparison here may be difficult, given the different experimental conditions present in the various investigations. Although differences in persistence are evident, there is no clear trend for more rapid degradation in the tropical soils versus the temperate ones that have been studied. Factors other than soil origin appear to be much more significant in modulating the rate of lindane dissipation. In comparison to flooded soils, lindane is much more persistent under nonflooded conditions. Yoshida and Castro (*48*) found that during a 1 month period very little lindane degraded in upland soils, whereas much of the added lindane was degraded in flooded soils. In flooded soils, more rapid degradation of lindane is associated with increased organic matter. Thus, Drego et al. (*49*) found that addition of green manure lowered lindane half-life in a flooded black clay from 12 to 5 days.

Table V. Degradation of lindane in flooded soils under laboratory conditions.

Soil Origin	DT_{50} (days)	Dose (ppm)	Soil and Conditions	Ref.
Philippines	ca. 25	15	clay, pH 4.7	50
Philippines	14-28	Unk	4 soils, rice fields	48
Japan	ca. 10	Unk	clay loam, rice field	51
India	ca. 50	Unk	acid sulfate, pH 3, 28% organic matter	52
	ca. 6		alluvial, pH 6.2	
	ca. 20		acid sulfate, pH 4.2	
	>120		sandy, pH 6, very low organic matter	
	ca. 15		laterite, pH 5	
USA	37	2	sandy loam, pH 6.4	53
India	ca. 15	1	sandy loam, pH 7.7, 0.8% organic carbon	54
India	ca. 12	1	black clay, pH 7.2	49
	ca. 5		black clay, pH 7.2, green manured	

Conclusions

The few available studies which have directly compared pesticide fate in temperate and tropical soils held under identical conditions (i.e., laboratory) reveal no significant differences in either the kinetics or pathway of degradation. It appears that there are no inherent differences in pesticide fate due to soil properties uniquely possessed by tropical soils. However, pesticides appear to dissipate significantly more rapidly under tropical conditions than under temperate conditions. The underlying mechanisms of this difference will be discussed in a succeeding section of this review.

How Does Pesticide Transport Under Tropical Conditions Compare With That for Temperate Conditions?

Compared to available data on pesticide transformation, much less data has been generated on the transport of pesticides in tropical ecosystems. Although the topics of pesticide sorption and leaching, surface runoff and erosion, and volatilization have been of interest for some time in the temperate zone, it has only been recently that broad interest has resulted in a significant number of investigations of these same topics under tropical conditions. As a result, it is not possible at this time to directly compare experimental data on pesticide transport and mobility under tropical and temperate conditions. However, a few examples will serve to illustrate some of the types of research that have been conducted in the tropics.

Interest in groundwater quality and the potential for pesticide contamination via leaching stimulated some early research on pesticide behavior in tropical environments. For example, Bovey et al. (55) compared the field leaching mobility of picloram in Texas and Puerto Rica soils, and found that under both conditions movement into deeper soil layers was an important avenue of dissipation for this product. More recent experiments have been focused on leaching mobility and potential groundwater entry of a number of products (56-60), and perhaps the most thorough of these investigations have been completed in connection with Hawaiian pineapple and sugarcane agriculture (61-63). At least some of these experiments, including those conducted in tropical island ecosystems (Hawaii, Sri Lanka, Barbados, Puerto Rico), have highlighted the potential vulnerability of poorly-sorbed pesticides to leaching and groundwater entry under conditions of high rainfall and shallow, permeable soils (58,62).

Runoff and surface water fate of pesticides under tropical conditions have only recently been the focus of investigation, and less information is available than for leaching behavior. Although a few surveys of surface water concentrations of pesticides in tropical areas have occurred, (64-65) few field-scale studies have been completed.

At this time few generalizations can be offered regarding the comparison of pesticide mobility in tropical versus temperate ecosystems. Clearly more research is needed prior to reaching any conclusions on the similarity or differences that may exist.

What Factors Help Explain Differences in Pesticide Behavior Under Tropical vs. Temperate Conditions?

In some instances, pesticides appear to dissipate significantly more rapidly under tropical conditions than under temperate conditions. The most prominent mechanisms for this acceleration appear to be related to the effect of tropical climates. These would include enhanced chemical degradation, biodegradation, and volatilization rates due to increased tropical temperatures, and accelerated photodegradation rates due to increased incident sunlight.

Interactions and Variability

One caveat to be offered regarding potential differences in pesticide dissipation rate between tropical and temperate conditions relates to the overall variability in pesticide behavior which may occur due to environmental factors. For example, in assessing pesticide behavior in soil, researchers are confronted with the tremendous degree of variability which results from the complex set of interactions involved. Laboratory degradation studies on the organophosphorus insecticide chlorpyrifos were conducted in 24 U.S. soils under similar laboratory conditions (1-10 ppm, darkness, % moisture holding moisture capacity). Observed degradation half-lives ranged from 10 to 325 days (*66*). Similarly, in 21 U.S. soils held under identical laboratory conditions, the sulfonamide herbicide flumetsulam displayed degradation half-lives of 13 to 130 days (*67*). A comparison of the variation in observed rates of soil degradation of 17 different pesticides yielded differences of 2X to 80X between minimum and maximum values for a given pesticide across the soil types examined (*68*). The above cited differences are due only to differences in soil properties, and the added variability contributed by environmental factors make comparisons difficult at best. This fact has important implications for comparisons of pesticide fate under tropical versus temperate conditions. For example, unless a sufficient diversity of temperate and tropical soil types are compared, it may not be possible to link any observed differences in pesticide degradation rate with the tropical or temperate origins of the soils. Instead, the differences observed may only be due to the variability one would expect when comparing different soil samples, regardless of their origin.

Increased Temperature

Hydrolysis

Hydrolytic degradation of pesticides in soil may occur due to reactions occurring in the soil pore water (e.g., base-catalyzed or acid-catalyzed) or on the surfaces of clay minerals (e.g., heterogeneous surface catalysis). Temperature has been considered a major factor modifying the rate of pesticide hydrolysis in water and soil. The acceleration of hydrolytic reactions has been generally well described by the Arrhenius Equation, which may be used to predict pesticide behavior in soil (69). An excellent example of the effect of temperature on pesticide hydrolysis in soil was provided by the work of Getzin (70), who investigated the fate of chlorpyrifos over the range 5 to 45°C and observed half-lives ranging from >20 to 1 day, respectively. Pure hydrolysis rates often increase by a factor of approximately 2X for each 10°C rise in temperature (69). Thus, it would be expected that in soil maintained at higher average year-round temperatures (i.e., tropical and subtropical regions), the rate of hydrolytic degradation would be considerably greater than in soil maintained at lower temperatures. The increase in rate, however, is highly dependent on the activation energy of the reaction. Soil pH has also been implicated as an important property influencing hydrolytic reactions of pesticide, although due to the complex nature of soil and operation of multiple hydrolytic mechanisms, construction of general principles has been lacking. The effects of soil pH on degradation of a given pesticide depend greatly on whether a compound is most susceptible to alkaline- or acid-catalyzed hydrolysis (71-73).

Although the relevance of hydrolytic degradation for soils in general and tropical soils in particular has not been well investigated, some work has been conducted on tropical soils. Korpraditskul et al. (25) investigated the chemical degradation of atrazine in a direct comparison of sterilized temperate and tropical soils. The study confirmed previous findings that abiotic, hydrolytic degradation was the prime loss mechanism for atrazine and that half-life was significantly correlated to soil pH. The authors concluded that at constant temperature and moisture hydrolytic degradation occurred more rapidly in lower pH soils, regardless of their origin (i.e., temperate vs. tropical). In addition, Korpraditskul et al. (22) demonstrated the dependency of atrazine hydrolytic degradation on temperature. After a 90 day incubation at 15, 25, 37, or 45°C, the percent atrazine remaining in a Thai soil was 70, 58, 41, and 27%, respectively. These data and others in the literature support the conclusion that chemical degradation of a pesticide through hydrolytic reactions is dependent on the nature of the chemical and the characteristics of the soil. These factors cannot be directly correlated to the region from which soils originate. However, the climate in which a soil is found can directly influence the rate of hydrolysis through modulation of the temperature and moisture of the soil.

Microbial Degradation

Soil microorganisms play an important role in the intermediate degradation and subsequent mineralization of many pesticides. Microbial degradation of a given pesticide may be of a cometabolic (i.e., incidental) nature, or may be linked with energy production or nutrient procurement supporting growth of the degrading population (74). An important consideration is the quite different microbially-mediated reactions which can be associated with aerobic or anaerobic conditions. Most investigations of soil microbial pesticide degradation in tropical soils have been associated with flooded, rice paddy conditions (45-46). Since soil microbial activities are strongly modulated by temperature, pesticide degradation would be expected to be greater in tropical soils, which experience higher year-round temperatures, than in temperate soils. This explanation would be consistent with observations of the elevated rates of soil organic matter turnover that characterize udic and ustic (rainy season) tropical environments (6). In an excellent review of microbial pesticide degradation in tropical soils, Sethunathan et al. (45) concluded that acceleration of microbial activities due to elevated temperatures was the major factor responsible for observations of increased degradation of pesticides under tropical rice paddy soil conditions. However, other environmental factors were also cited as potentially important variables governing microbial activities. In many tropical areas characterized by intermittent heavy rain and dry seasons, soils are subjected to alternate periods of flooding and drying with concomitant increases in the activities of anaerobic and aerobic microorganisms, respectively. The authors felt that such alternate reduction and oxidation cycles in the soil could provided a favorable environment for more extensive destruction of organic compounds than in either system alone. For example, diazinon was readily cleaved via hydrolysis in flooded soils, but complete mineralization of the resulting aromatic-ring metabolite only occurred under aerobic conditions following anaerobiosis (75).

Enhanced biodegradation is a phenomenon whereby a pesticide is rapidly degraded (i.e., catabolized) by soil microorganisms which have adapted due to a previous exposure to the pesticide (74). Pesticides proven to be susceptible to enhanced biodegradation include aldicarb, carbaryl, carbofuran, 1,3-D, 2,4-D, EPTC, fenamiphos, isofenphos, and terbufos. Although adapted microbial degradation has been most heavily studied in soils from North America and Europe (74,76-78), enhanced biodegradation of several of these pesticides (carbaryl, carbofuran, 1,3-D, diazinon, terbufos) in soils of tropical and subtropical origins has been documented (45,79-82). Thus, microbial adaptation for enhanced pesticide biodegradation appears to be a globally occurring phenomenon, and it is possible that similar microbial strains, enzymes, or genes may be involved. Some evidence of this is provided by investigations of bacteria involved in the degradation of parathion. Biochemical examination of a *Flavobacterium* sp. isolated in the Philippines and a *Pseudomonas* sp. isolated in the USA revealed that the plasmid-encoded genes for parathion hydrolase production were nearly identical, indicating a common homology (83-85).

Volatility

The loss of a chemical from the surface of the soil through volatilization is influenced by the physico-chemical properties of the chemical, the method of application, the properties of the soil (e.g., temperature, moisture), the concentration of the chemical, and the weather conditions. As previously stated, tropical soils cannot necessarily be classified as a distinct entity with a unique set of properties. Indeed, soils from within any single continent, territory, or region may vary significantly. Based on this observation, the major factor affecting the amount of volatilization from the soil in tropical climates will be the prevailing environmental conditions. Several researchers have demonstrated a roughly 3-4X increase in volatility for each 10°C rise in temperature (*86-88*). The effects of relative humidity and temperature on the loss of alachlor from a soil surface were investigated by Hargrove and Merkle (*89*). Results showed that alachlor loss from the soil increased with increasing temperature and relative humidity, with humidity having the greatest impact on volatilization at higher (38°C) rather than lower (20°C) temperature. The findings of the various investigations related to variables affecting volatility suggest that the major influence for any particular pesticide will be climatic conditions such as temperature, soil moisture, relative humidity (insofar as it influences soil moisture), and wind turbulence. Few investigations of pesticide volatility have been carried out under tropical soil conditions. In a few instances, tropical pesticide fate researchers have reported that increased dissipation under field conditions, as compared with published results from temperate regions, appeared to be related to increased volatilization. For example, a coordinated field and laboratory program on the fate of DDT in tropical soils concluded that the more rapid dissipation of this persistent compound in the tropics was largely due to increased volatility under tropical conditions (*24*).

Increased Sunlight

Photodegradation

In the past it was commonly assumed that photolytic degradation was not an important mechanism of pesticide loss from soil. Recent evidence increasingly suggests that photoinduced transformations can, in some instances, be significant. Although a pesticide may not be directly transformed by solar radiation, due to low absorbance between 290 and 400 nm wavelengths, indirect photodegradation may still be an important factor. Gohre and Miller (*90*) demonstrated that photoinduced oxidizing species (e.g., singlet oxygen, peroxide) are produced when soil is exposed to sunlight. From the results of this study, organic fractions of the soil were postulated as being the sensitizing species. In a separate study, focused on photooxidation of parathion to paraoxon on soil, Spencer et al. (*91*) demonstrated that the type of clay mineral in the soil was the dominant feature in catalyzing the oxidation reaction (i.e., kaolinite >> montmorillonite). The major factors affecting oxidation of parathion were concluded to be atmospheric ozone concentration, UV light, and the nature of the soil, with soil organic content being inversely related to rate of oxidation. The significance of solar induced transformations was illustrated by the work of Zayed et al. (*92*), who reported that the degradation of DDT (primarily to DDE) in soil was enhanced by exposure to sunlight. Over a 90-day period of exposure, only 65% of the initial DDT remained compared to 91% in the unexposed, dark control. In contrast to photosensitization, work by Miller and Zepp (93) reported on the apparent quenching effects of humic and mineral materials in soil. Several researchers have reported significantly more rapid photodegradation of pesticides on moist soil surfaces versus dry soil surfaces (94-95). Regarding the effect of soil type on photolysis, little information of a predictive nature has been generated, and therefore the intensity and spectral distribution of the solar radiation and possibly moisture status of the soil should be considered the major predictive factors.

Given that sunlight intensity can be a major factor governing rates of soil photolysis of pesticides, variations due to geographical location and season would be expected. Although estimation of soil photolysis as influenced by these factors has not been directly investigated, the kinetics of photolysis in aqueous systems has received further attention. Based on the more uniform light intensities throughout the year in the tropics, quantum yield calculations predict that half-lives of photosensitive pesticides in the tropics will be generally shorter than those observed in temperate regions (Figure 3).

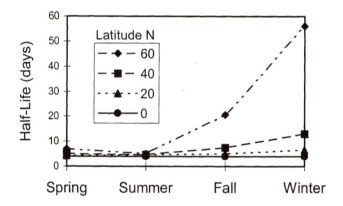

Figure 3. Seasonal and latitudinal dependence of fenvalerate photodegradation (adapted from Reference 96).

Conclusions

The availability of information on the ecosystem fate of pesticides under tropical conditions is somewhat limited as compared to that from temperate ecosystems. This is especially true for studies of pesticide mobility, and somewhat less so for studies of pesticide degradation and dissipation. Studies which have directly compared pesticide fate in temperate and tropical soils and waters held under identical conditions (i.e., laboratory), reveal no significant differences in either the kinetics or pathway of degradation. Thus, it appears that there are no inherent differences in pesticide fate due to properties uniquely possessed by tropical environmental matrices. Additionally, mechanistic data from temperate zone investigations would appear to be equally applicable for tropical zone conditions. Pesticides appear to dissipate significantly more rapidly under tropical ecosystem conditions than under temperate conditions. The most prominent mechanisms for this acceleration in pesticide dissipation appear to be related to the effect of tropical climates. These would include enhanced chemical degradation, biodegradation, and volatilization rates due to increased temperatures and increased photodegradation rates due to greater levels of incident sunlight.

There are important implications of these conclusions regarding key concerns related to pesticide use in the tropics (efficacy, environmental quality, human safety). First, laboratory data on pesticide transformation behavior in temperate soils is clearly predictive of the pathway and mechanism of degradation in tropical soils. Thus, provided that these studies have examined a suitable range of conditions, special studies on tropical soils or waters would appear to be unwarranted. Second, pesticide dissipation rates obtained under temperate field conditions may not be representative of those under tropical field conditions. Based on the observed increase in dissipation rate under many tropical conditions, the employment of temperate zone dissipation rates for tropical ecosystem fate assessments may result in overly conservative conclusions. Finally, too little direct data on pesticide transport under tropical conditions exists to allow meaningful comparisons with that observed under temperate conditions. Given recent advances in the field of environmental simulation modeling, this would appear to be a fertile area for further model construction and field validation under tropical conditions.

Investigations on the fate of pesticides in tropical ecosystems should continue to be encouraged, especially as they relate to leaching, runoff, and volatilization mobility. The inordinate focus of research on pesticide fate in temperate ecosystems should be balanced by increased efforts in tropical ecosystems. Rather than overwhelm the scientific literature with data obtained exclusively under temperate conditions (e.g., another atrazine degradation or leaching study in a U.S. Corn Belt soil), academic, government, and agrochemical industry researchers should consider incorporating tropical systems and their issues into testing programs as appropriate. In addition, pesticide regulatory agencies for countries with significant tropical area should encourage field validation and/or modeling rather than require additional laboratory studies as a means of obtaining the most useful and regionally-specific information on pesticide fate in tropical ecosystems. Further attempts should be made to validate environmental fate models for application to simulation of pesticide dissipation and especially mobility under tropical conditions.

120

References

1. Mauldin, J.; Jones S.; Beal, R. *Pest Control* 1987, *55*, 46-59.
2. Lenz, M.; Watson, J. A. L.; Barrett, R. A. *Australian Efficacy Data for Chemicals Used in Soil Barriers Against Subterranean Termites*, CSIRO, Division of Entomology Technical Paper, **1988**, *27*, 1-23.
3. Sornnuwat, Y.; Vongkaluang, C.; Takahashi, M.; Yoshimura, T. *Mokuzai Gakkaishi* **1996**, *42*, 520-531.
4. Jain, V. *Environ. Sci. Technol.*, **1992**, *26*, 226-228.
5. Jensen, J. K. *Pestic. Outlook*, **1992**, *3*, 30-33.
6. Sanchez, P. A. *Properties and Management of Soils in the Tropics*, John Wiley & Sons, New York, 1976.
7. Cobley, L. S.; Steele, W. M. *An Introduction to the Botany of Tropical Crops*, Longman, New York, 1984.
8. Uehara, G.; Gillman, G. *The Mineralogy, Chemistry, and Physics of Tropical Soils with Variable Charge Clays*, Westview Press, Boulder, Colorado,1981.
9. Isbell, R. F. *Proceedings of the International Workshop on Soils*, Townsville, Queensland, Australia, 1983, p. 17.
10. Soil Survey Staff, *Soil Taxonomy*, U.S. Department of Agriculture, Handbook 436, Washington, DC, 1975.
11. Buol, S. W.; Hole, F. D.; McCracken, R. J. *Soil Genesis and Classification*, Iowa State University Press, Ames, Iowa, 1989.
12. Aubert, G.; Tavernier, R.; In: *Soils of the Humid Tropics*, National Academy of Sciences, Washington, DC, 1972, pp. 17-44.
13. Tarrant, J. (ed.) *Farming and Food*, Oxford University Press, New York, 1991.
14. Phillips, M.; McDougall, J.; Mathisen, F.; Galloway, F. *Agrochemical Country Markets*, Wood Mackenzie, London, 1998.
15. Mellor, J. W.; Adams, R. H. *Chem. Eng. News*, **1984**, April 23, pp. 32-39.
16. Pimental, D. *Estimated Annual World Pesticide Use*, Ford Foundation, Facts and Figures, New York, 1990.
17. Logan, J. W. M.; Buckley, D. S. *Pestic. Outlook*, **1991**, pp. 33-37.
18. Gratz, N. G. *Ann. Rev. Entomol.*, **1999**, *44*, 51-75.
19. Chavasse, D. C.; Yapp, H. H. *Chemical Methods for the Control of Vectors and Pests of Public Health Importance*, World Health Organization, WHO/CTD/WHOPES/97.2, 1997.
20. Racke, K. D.; Skidmore, M. W.; Hamilton, D. J.; Unsworth, J. B.; Miyamoto, J.; Cohen, S. Z. *Pure Appl. Chem.*, **1997**, *69*, 1349-1371.
21. Simon, L.; Spiteller, M.; Wallnöfer, P. R. *J. Agric. Food Chem.*, **1992**, *40*, 312-317.
22. Korpraditskul, R.; Korpraditskul, V.; Kuwatsuka, S. *J. Pestic. Sci.*, **1992**, *17*, 287-289.

23. Walker, A.; Hance, R. J.; Allen, J. G.; Briggs, G. G.; Chen, Y. L.; Gaynor, J. D.; Hogue, E. J.; Malquori, A.; Moody, K.; Moyer, J. R.; Pestemer, W.; Rahman, A.; Smith, A. E.; Streibig, J. C.; Torstensson, N. T. L.; Widyanto, L. S.; Zandvoort, R. *Weed Res.*, **1993**, *23*, 373-383.
24. Hassan, A. *J. Environ. Sci. Health*, **1994**, *B29*, 205-226.
25. Korpraditskul, R.; Katayama, A.; Kuwatsuka, S.; *J. Pestic. Sci.*, **1993**, *18*, 77-83.
26. Walker, A.; Barnes, A. *Pestic. Sci.*, **1981**, *12*, 123-132.
27. IAEA, *Isotope Techniques for Studying the Fate of Persistent Pesticides in the Tropics*, , International Atomic Energy Agency, TECDOC-476, Vienna, 1988.
28. Lichtenstein, E. P.; Schulz, K. R. *J. Econ. Entomol.*, **1959**, *52*, 118-124.
29. Edwards, C. A. *Resid. Rev.* **1966**, *13*, 83-132.
30. Nash, R. G.; Woolson, E. A. *Science*, **1967**, *157*, 924-927.
31. Ray, S. M.; Trieff, N. M. In: *Environment and Health*, Trieff, N. M. (ed.), Ann Arbor Science, Ann Arbor, Michigan, 1980, pp. 93-120.
32. Woodwell, G. M.; Craig, P. P.; Johnson, H. A. *Science*, **1971**, *174*, 1101-1107.
33. Gaeb, S.; Nitz, S.; Parlor, H.; Korte, F. *Chemosphere*, **1975**, *4*, 251.
34. Billings, W. N.; Bidleman, T. F. *Atmosph. Environ.* **1983**, *17*, 383.
35. Lalah, L. O.; Acholla, F. V.; Wandiga, S. O. *J. Environ. Sci. Health*, **1994**, *B29*, 57-64.
36. Stephens, J.; Maeda, D. N.; Ngowi, A. V.; Moshi, A. O.; Mushy, P.; Mausa, E. *J. Environ. Sci. Health*, **1994**, *B29*, 65-71.
37. Agarwhal, H. C.; Singh, D. K.; Sharma, V. B. *J. Environ. Sci. Health*, **1994**, *B29*, 189-194.
38. Hussain, A.; Maqbool, U.; Asi, M. *J. Environ. Sci. Health*, **1994**, *B29*, 1-15.
39. Sjoeib, F.; Anwar, E.; Tungguldihardjo, M. S. *J. Environ. Sci. Health*, **1994**, *B29*, 17-24.
40. Varca, L. M.; Magllona, E. D. *J. Environ. Sci. Health*, **1994**, *B29*, 25-35.
41. Xu, B.; Jianying, G.; Yongxi, Z.; Haibo, L. *J. Environ. Sci. Health*, 1994, **1994**, *B29*, 37-46.
42. Espinosa-Gonzalez, J.; Garcia, V.; Ceballos, J. *J. Environ. Sci. Health*, **1994**, *B29*, 97-102.
43. Andrea, M. M.; Luchini, L. C.; Mello, M. H .S. H.; Tomita, R. Y.; Mesquita, T. B.; Musumeci, M. R. *J. Environ. Sci. Health*, **1994**, *B29*, 121-132.
44. Helling, C. S.; Engelke, B. F.; Doherty, M. A. *J. Environ. Sci. Health*, **1994**, *B29*, 103-119.
45. Sethunathan, N.; Adhya, T. K.; Raghu, K. In *Biodegradation of Pesticides*, F. Matsumura and C. R. Krishna-Murty, (eds.), Plenum Press, New York, 1982, pp. 91-115.
46. Yoshida, T.; In *Soil Biochemistry, Vol. 3*, Paul, E. A.; McLaren, A. D. (eds.), 83-117, Marcel Dekker, New York 1975, pp. 83-117.

47. Abdullah, A. R.; Bajet, C. M.; Matin, M. A.; Nhan, D. D.; Sulaiman, A. H. *Environ. Toxicol. Chem.*, **1997**, *16*, 59-70.

48. Yoshida, T.; Castro, T. F. *Soil Sci. Soc. Am. Proc.*, **1970**, *34*, 440-442.

49. Drego, J. D.; Murthy, N. B. K.; Raghu, K. *J. Agric. Food Chem.* **1990**, *38*, 266-268.

50. Macrae, I. C.; Raghu , K.; Castro, T. F. *J. Agric. Food Chem.* **1967**, *15*, 911-914.

51. Tsukano, Y. *Jap. Agric. Res. Quart.* **1973**, *7*, 93-97.

52. Siddaramapa, R.; Sethunathan, N. *Pestic. Sci.*, **1975**, *6*, 395-403.

53. Jordan, E. G. *Metabolism of Lindane in Soil Under Aerobic and Anaerobic Conditions*, unpublished report of Rhone-Poulenc, 1988.

54. Samual, T.; Pillai, M. K. *Archiv. Environ. Contam. Toxicol.* **1990**, *19*, 214-220.

55. Bovey, R. W.; Dowler, C. C.; Merkle, M. G. *Pestic. Monit. J.*, **1969**, *3*, 177-181.

56. Agnihotri, N. P.; Pandey, S. Y.; Jain ,H. K.; Srivastava, K. P. *Ind. J. Agric. Chem.*, **1981**, *14*, 27-31.

57. Akinyemiju, O. A. *Proc. Brighton Crop Protec. Conf.* **1991**, pp. 485-490.

58. Chilton, P. J.; Lawrence, A. R.; Barker, J. A. In: *Hydrological, Chemical, and Biological Processes of Transformation and Transport of Contaminants in Aquatic Environments*, IAHS Publication 219, British Geological Survey, Wallingford, Oxford, 1994, pp. 51-66.

59. Mohapatra, S. P.; Agnihotri, N. P. *Pestic. Outlook*, **1996**, *7*, 27-30.

60. Laabs, V. Amelung, W.; Pinto, A.; Alstaedt, A.; Zech, W. *Chemosphere*, **1999**, in press.

61. Lee, C. C.; Green, R .E.; Apt, W. J. *J. Contam. Hydrol.*, **1986**, *1*, 211-225.

62. Oki, D. S.; Giambelluca, T. W. *Water Resources Bull.*, **1989**, *25*, 285-294.

63. Loague, K.; Miyahira, R. N.; Oki, D. S.; Green, R. E.; Schneider, R. C.; Giambelluca, T. W. *Ground Water*, **1994**, *32*, 986-996.

64. Castillo, L. E.; De La Cruz, E.; Ruepert, C. *Environ. Toxicol. Chem.*, **1997**, *16*, 41-51.

65. Simpson, B.; Kwong, R. N. K.; Hargreaves, P. *9th IUPAC International Congress of Pesticide Chemistry*, London, Abstract 6C-017, 1998.

66. Racke, K. D.; Laskowski, D. A.; Schultz, M. R. *J. Agric. Food Chem.*, **1990**, *38*, 1430-1436.

67. Lehman, R. G.; Miller, J. R.; Fontaine, D. D.; Laskowski, D. A.; Hunter, J. H.; Cordes, R. C. *Weed Res.*, **1992**, *32*, 197-205.

68. Laskowski, D. A. In *Agrochemical Environmental Fate: State of the Art*, Leng, M. L.; Leovey, E. M. K.; Zubkoff, P. L. (eds.), Lewis Publishers, Boca Raton, Florida 1995, pp. 117-128.

69. Laskowski, D. A. *Resid. Rev.*, **1983**, 85, 139-147.

70. Getzin, L. W. *J. Econ. Entomol.*, **1981**, *74*, 707-713.

71. Armstrong, R. E.; Chesters, G.; Harris, R. F. *Soil Sci. Soc. Am. Proc.*, **1967**, *31*, 61-66.

72. Racke, K. D.; Steele, K. P.; Yoder, R. N.; Dick, W. A.; Avidov, E. *J. Agric. Food Chem.*, **1986**, *44*, 1582-1592.
73. Guth, J. A. In: *Interactions Between Herbicides and the Soil*, Hance, R. J. (ed.), Academic Press, London, 1980.
74. Racke, K. D.; Coats, J. R. (eds.), *Enhanced Biodegradation of Pesticides in the Environment*, Symposium Series No. 426, American Chemical Society, Washington, DC, 1990.
75. Sethunathan, N. *Adv. Chem. Ser.* **1972**, *111*, 244.
76. Roeth, F. W. *Rev. Weed Sci.*, **1986**, *2*, 45-66.
77. Suett, D. L.; Walker, D. L. *Aspects App. Biol.*, **1988**, *17*, 213-222.
78. Felsot, A. S. *Ann. Rev. Entomol.*, **1989**, *34*, 453-476.
79. Sethunathan, N.; Pathak , M. D. *J. Agric. Food Chem.* **1972**, *20*, 586-589.
80. Ramanand, K.; Panda, S.; Sharmila, M.; Adhya, T. K.; Sethunathan, N. *J. Agric. Food Chem.*, **1988**, *36*, 200-205.
81. Wolt, J. D.; Holbrook, D. L.; Batzer, F. R.; Balcer, J. L.; Peterson, J. R. *Acta Hortic.*, **1993**, *334*, 361-371.
82. Felsot, A. S. American Chemical Society, San Diego, CA, Abstract AGRO 84, 1994.
83. Mulbry, W. W.; Karns, J. S. *J. Bacteriol.*, **1989**, *171*, 6740-6746.
84. Serdar, C. M.; Murdock, D. C.; Rhode, M. F. *Biotechnol.*, **1989**, *7*, 1151-1155.
85. Head, I. M.; Cain, R. B. *Pestic. Outlook*, **1991**, *2*, 13-16.
86. Burkhard, N.; Guth, J. A. *Pestic. Sci.*, **1981**, *12*, 37-44.
87. Farmer, W. J.; Igue, K.; Spencer, W. F.; Martin, J. P. *Soil Sci. Soc. Am. Proc.* **1972**, *36*, 443-447.
88. Grover, R. *Weed Sci.*, **1975**, *23*, 529.
89. Hargrove, S.; Merkle, M. G. *Weed Sci.*, **1971**, *19*, 652-654.
90. Gohre, K.; Miller, G. C. *J. Agric. Food Chem.*, **1986**, *31*, 1164-1168.
91. Spencer, W. F.; Adams, J. D.; Shoup, D.; Spear, R. C. *J. Agric. Food Chem.*, **1980**, *28*, 366-371.
92. Zayed, S. M. A. D.; Mostafa, I. Y.; El-Arab, A. E. *J. Environ. Sci. Health*, **1994**, *B29*, 47-56.
93. Miller, G. C.; Zepp, R. G. *Residue Rev.*, **1983**, *85*, 89-125.
94. Burkhard, N.; Guth, J. A. *Pestic. Sci.*, **1979**, *10*, 313-319.
95. Rainey, D. P.; O'Neill, J. D.; Saunders, D. G.; Powers, F. L.; Zabik, J. M.; Babbitt, G. E. American Chemical Society, Abstract AGRO 046, New Orleans, Lousiana, 1995.
96. Miyamoto, J.; In *Assessment and Management of Risks from Pesticides Use in SE Asia*, Bangkok, E1, 1992.

Chapter 7

Simple Determination of Herbicides in Rice Paddy Water by Immunoassay

Shiro Miyake[1,4,6], Yasuo Ishii[2], Yuki Yamaguchi[1], Katsuya Ohde[1],
Minoru Motoki[3], Mitsuyasu Kawata[3], Shigekazu Ito[4,6], Yojiro Yuasa[4],
and Hideo Ohkawa[5]

[1]Narita R&D Center, Iatron Laboratory Inc., Katori-gun, Chiba 289–2247, Japan
[2]Division of Pesticide, National Institute of Agro-Environmental Sciences,
Tsukuba-shi, Ibaraki 305–8604, Japan
[3]Naruto Research Center, Otsuka Chemical Company, Naruto-shi,
Tokushima 772–0022, Japan
[4]Environmental Immuno-Chemical Technology Company, Minato-ku,
Tokyo 105–0013, Japan
[5]Research Center for Environmental Genomics, Kobe University, Kobe-shi,
Hyogo 657–8501, Japan
[6]Current address: Bio Applied Systems Inc., Kyoto-shi, Kyoto 601–8305, Japan

Studies were conducted on a simple and rapid analytical
method for the determination of herbicide residues in paddy
field water using an immunoassay technique. For the studies,
two sulfonylurea herbicides, bensulfuron-methyl and pyrazosulfuron-ethyl, and a carbamate herbicide, thiobencarb, which
are widely being used on rice paddies in Japan, were selected
as representative paddy field herbicides or components of
herbicide mixtures. A method for simple, rapid, simultaneous
immunoassay for the three herbicide residues in water was
established as follows. Highly reactive monoclonal antibodies
(MoAbs), which selectively react with corresponding
herbicides, were prepared for the three herbicides. Then, the
procedure for a direct competitive enzyme-linked
immunosorbent assay (direct C-ELISA) was established by
utilizing MoAbs in order to achieve a simultaneous
determination of the three herbicides. By this analytical
method, each of the three herbicide residues was determined

simultaneously in paddy water samples from different locations at a lower limit of 0.2 ng/ml, without any pre-treatment. Analytical data obtained by the direct C-ELISA were shown to possess good correlation with those obtained by chemical (instrumental) analysis. Thus, the simple simultaneous C-ELISA was successfully developed for the three herbicides, for which up to 100 or more water samples can be analyzed in only 2 to 3 hours.

Introduction

Two sulfonylurea herbicides, bensulfuron-methyl and pyrazosulfuron-ethyl, and a carbamate herbicide, thiobencarb, whose structures are shown in Figure 1, are widely used for control of weeds on rice paddies in Japan as a single herbicide or a component of herbicide mixtures. Although the rice paddies are usually well managed, paddy waters containing these herbicides may run off from the paddies. The water pollution by the herbicides may bring potential hazards to various kinds of plants in the area and even to human health. Therefore, the herbicide residues in the polluted water should be quantitatively monitored.

In general, residue analysis of herbicides is carried out by using chemical analytical methodology, such as gas chromatography (GC) or high performance liquid chromatography (HPLC). These analytical procedures are sensitive and precise, but labor-intensive and time consuming, because a large number of water samples from the environment must be determined over a short time period. Therefore, a more convenient method has been desired.

A direct competitive enzyme-linked immunosorbent assay (direct C-ELISA) is known as a simple assay method for determination of herbicide residues in water (1-4). This study is concerned with the preparation of monoclonal antibodies for the three different herbicides, bensulfuron-methyl, pyrazosulfuron-ethyl and thiobencarb, and the development of direct C-ELISAs, based on these antibodies, which enable simultaneous determination of these herbicides, in paddy waters.

Materials and Methods

Preparation of Polyclonal and Monoclonal Antibodies

The carboxylated derivatives of the three herbicides (haptens), whose structures are shown in Figure 1, were conjugated to keyhole limpet hemocyanin (Pierce Chemical) as described previously (5). Each conjugate was injected

126

into Balb/c mice (Clea Japan, Japan). Seven days after the boost immunization, anti-serum was prepared from samples of mouse blood and used as polyclonal antibodies for bensulfuron-methyl, pyrazosulfuron-ethyl or thiobencarb. Fourteen days after the boost immunization, the final immunization was carried out. Spleen cells of each mouse were used for monoclonal antibody preparation as described previously (5). After cell fusion, the hybridomas grown in the wells of 96-well microtiter plates were screened by comparison of the reactivities between the monoclonal antibody produced and the corresponding polyclonal antibody. The monoclonal antibodies for the three herbicides were purified respectively, from the cultured fluids of the isolated hybridomas through precipitation by final 50% of saturated ammonium sulfate added, dissolution with 0.2 volume of distilled water and re-precipitation by final 35% of saturated ammonium sulfate added.

Indirect C-ELISA

Each indirect C-ELISA for bensulfuron-methyl, pyrazosulfuron-ethyl and thiobencarb was carried out as described previously (5). In brief, the wells of microtiter plate with 96-wells (Corning) were respectively coated with the conjugate of each hapten and bovine serum albumin (BSA, Sigma Chemicals), prepared by binding of carboxylic acid moiety of the hapten and amino group of the BSA using conventional active ester method (5), to saturate the well surface. The solution containing one of the three herbicides and 1% methanol was added to the corresponding wells and immediately followed by adding the corresponding monoclonal antibody solution. The wells were incubated for 1 hr at 25°C and then washed 3 times. After the solution of anti-mouse IgG antibody conjugated with horseradish peroxidase (ICN Pharmaceuticals) was added, the wells were incubated again for 1 hr at 25°C. After the wells were washed again, the color was developed in the wells by using 3,3',5,5'-tetramethylbenzidine (Dojindo Laboratories, Japan) as a substrate of horseradish peroxidase (HRP). After 10 min, the color development was terminated by adding 1 M sulfuric acid and the absorbance was measured at 450 nm on an microplate reader (Bio-Tek Instruments).

Direct C-ELISA for Simultaneous Determination of three Herbicides (Simultaneous C-ELISA)

Each separate direct C-ELISA system was initially developed for determination of bensulfuron-methyl, pyrazosulfuron-ethyl and thiobencarb, as described previously (6). In brief, each specific monoclonal antibody was separately coated on the wells of 96-well microtiter plates. One ml each of water sample containing bensulfuron-methyl, pyrazosulfuron-ethyl or thiobencarb was mixed with equal volume of the corresponding conjugate of the hapten and HRP

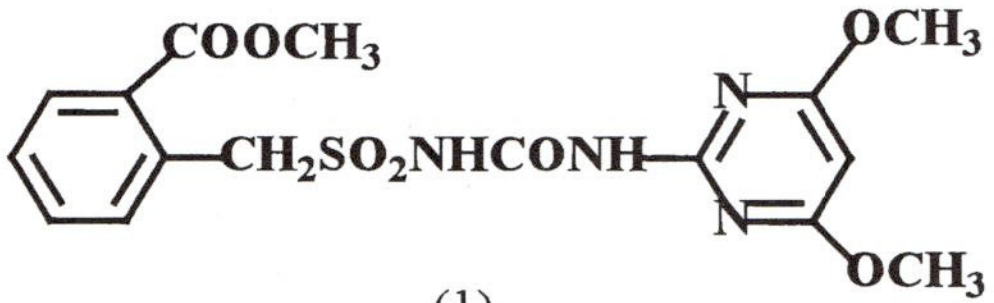

(1)

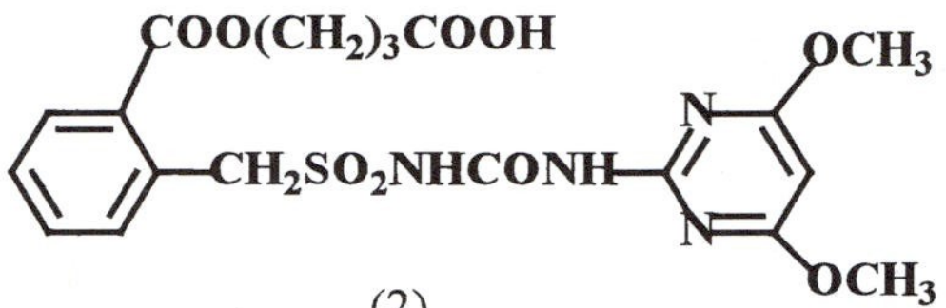

(2)

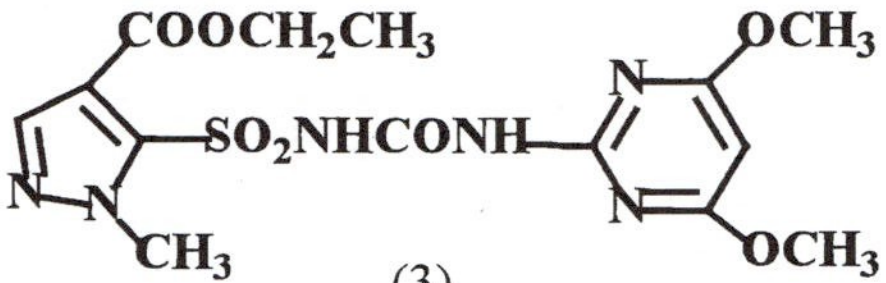

(3)

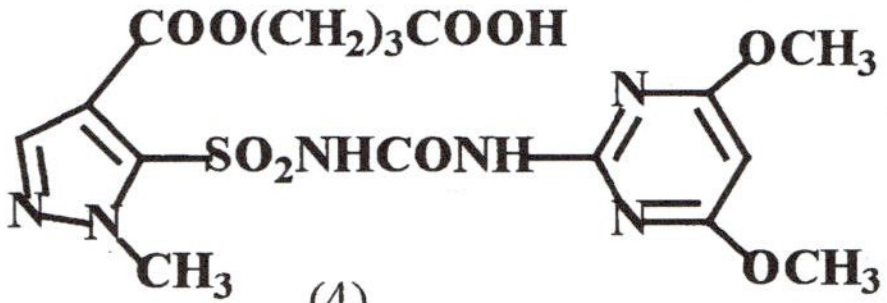

(4)

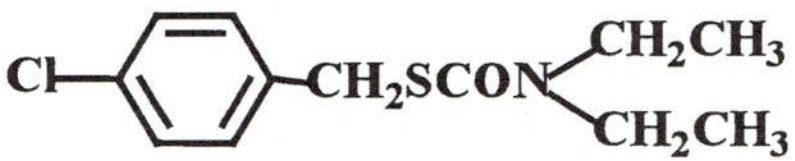

(5)

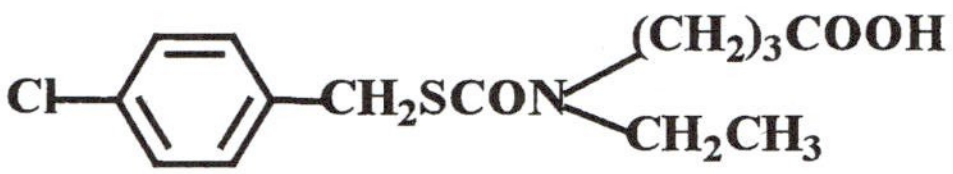

(6)

*Figure 1. Structure of bensulfuron-methyl (**1**), bensulfuron-methyl hapten (**2**), pyrazosulfuron-ethyl (**3**), pyrazosulfuron-ethyl hapten (**4**), thiobencarb (**5**) and thiobencarb hapten (**6**).*

(HRP-hapten) prepared by using conventional active ester method as well as the conjugate of the hapten and BSA. The HRP was purchased from Toyobo, Japan. Then, 100 μl aliquots of the mixture of the sample and the HRP-hapten were added to the wells of the corresponding microtiter plate. After incubation for 30 min at 25° C, the wells were washed 3 times, and the color was developed as described above.

The simultaneous C-ELISA was performed combining the three HRP-haptens into a mixture, which was then mixed with an equal volume of water sample containing bensulfuron-methyl, pyrazosulfuron-ethyl and/or thiobencarb, as shown in Figure 2. On the other hand, the wells of a 96-well microtiter plate were separately coated with each of monoclonal antibodies for the three herbicides, for example the wells of line B and C of the microtiter plate were coated with the monoclonal antibody for bensulfuron-methyl, line D and E was for pyrazosulufuron-ethyl and line F and G was for thiobencarb. Then, 100 μl aliquots of the mixture of the water sample and the three HRP-haptens were added to the corresponding wells. Incubation of the plate, washing and color development were carried out as described above.

Instrumental Analyses

Extraction Procedures were carried out on the Sep-Pak PS-2 cartridges (Waters) prepared by washing with 5 ml each of ethyl acetate, methanol and water. Water samples were filtered through 1 μm filters (Whatman GF/B) and adjusted to pH 3 with diluted phosphoric acid. A five hundreds ml of sample water was passed through the Sep-Pak PS-2 cartridge using a Millipore Workstation (Waters) at 10 ml/min. The cartridge was dried with air after loading and then eluted with 5 ml of ethyl acetate. The ethyl acetate eluent was dehydrated with sodium sulfate anhydride and evaporated to dryness. The residue was dissolved in 5 ml acetone for GC determination of thiobencarb or 5 ml acetonitrile/water (1:1, v/v) for HPLC/MS/MS determination of bensulfuron-methyl and pyrazosulfuron-ethyl.

GC Analysis was carried out on a Shimadzu Model GC-17A equipped with a flame thermionic detector operating on the nitrogen mode. The injection volume onto GC was 2 μl. Operating conditions were as follows: column, DB-5 (J&W), 0.32 mm I.D. x 30 m, film thickness 250 μm ; inlet temperature 280 ° C; column temperature, initially held at 60 ° C for 3 min, then, 10 ° C /min to 280 ° C and held at 280 ° C for 3 min; detector temperature, 290 ° C; carrier gas, He,

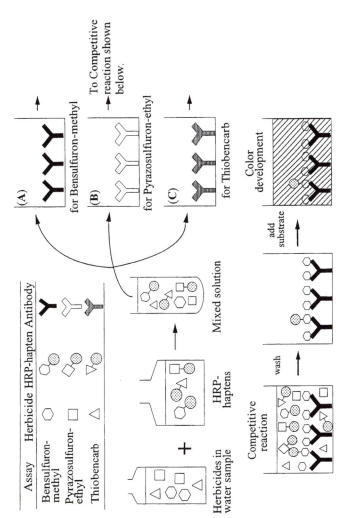

Figure 2. Flow chart for simultaneous C-ELISA (A), (B) and (C) represent single micro-wells of the different position on a 96-well microtiter plate.

initially held at 110 kPa for 1 min, then, -60 kPa/min to 50 kPa, held at 50 kPa for 1 min, then, 4 kPa/min to 138 kPa and held at 138 kPa for 3 min; make up gas, He 30 ml/min; hydrogen, 1 ml/min; air 225 ml/min. Recovery from test samples fortified with 1-10 ng/ml of thiobencarb averaged >90%.

HPLC/MS/MS Analysis was carried out on a Hewlett-Packard Model 1090 and a Perkin-Elmer Model API 300 triple quadrupole mass spectrometer equipped with a TurboIonSpray^{TM} electrospray ionization inlet. The injection volume onto HPLC was 10 µl. HPLC operating conditions were as follows: column, Capcell Pak C 18 UG120, 1.5 mm I.D. x 35 mm (Shiseido, Japan); guard column, OPTI-GUARD C18, 1 mm I.D.x15 mm (Optimize Technologies Inc.); oven temperature, 40 °C; mobile phase, acetonitrile/water (1:1, v/v) containing 20 mM acetic acid; flow rate, 0.06 ml/min. Mass spectrometer operating conditions: measurement, selected ion monitoring (SRM) mode; collisionlly activated dissociation (CAD) gas, nitrogen; heated gas temperature, 400 °C; monitored MS/MS trasition, bensulfuron-methyl m/z 411([M+H]+) - 149, pyrazosuslfuron-ethyl m/z 425 ([M+H]+) - 182. Recovery from test samples fortified with 1 - 10 ng/ml of bensulfuron-methyl and pyrazosulfuron-ethyl averaged >90%.

Results and Discussion

Preparation of Monoclonal Antibodies and their Reactivities

A number of hybridomas secreting a monoclonal antibody which reacted to the corresponding herbicide were found during hybridoma screening. However, most of the hybridomas were useless because the binding ability of their monoclonal antibodies were similar to that of the polyclonal antibody from the same mouse. Through the screening, the monoclonal antibody OC-1/519-4 was selected as the most reactive one for bensulfuron-methyl. Similarly, the monoclonal antibody OC-2/14-3 was selected for pyrazosulfuron-ethyl and TBC7-2 for thiobencarb.

The reactivity of the three monoclonal antibodies with the corresponding herbicides was compared to that of the polyclonal antibodies on indirect C-ELISA, as shown in Figure 3. The 50% inhibition concentration (I_{50}) values were 2 ng/ml and 2000 ng/ml for OC-1/519-4 and the corresponding polyclonal antibody, 20 ng/ml and 700 ng/ml for OC-2/14-3 and the polyclonal antibody, and 2 ng/ml and 80 ng/ml for TBC7-2 and the polyclonal antibody, respectively. As expected from the screening for the most reactive monoclonal antibody, reactivity of each monoclonal antibody was higher than the corresponding polyclonal

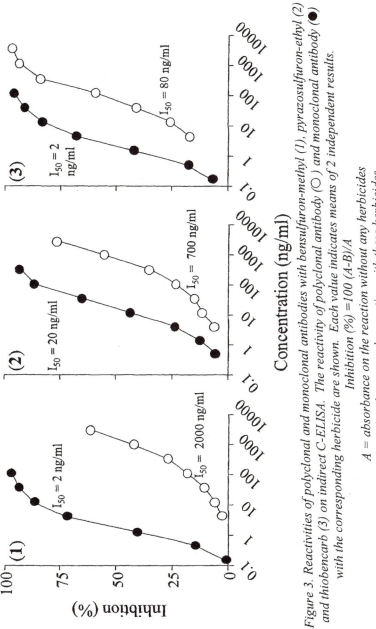

Figure 3. Reactivities of polyclonal and monoclonal antibodies with bensulfuron-methyl (1), pyrazosulfuron-ethyl (2) and thiobencarb (3) on indirect C-ELISA. The reactivity of polyclonal antibody (○) and monoclonal antibody (●) are shown. Each value indicates means of 2 independent results.

Inhibition (%) = 100 (A-B)/A

A = absorbance on the reaction without any herbicides

B = absorbance on the reaction with these herbicides

antibody. In particular, reactivity of OC-1/519-4 with bensulfuron-methyl was 1000-fold higher than the polyclonal antibody.

Cross-Reactivities of Monoclonal Antibodies with Related Herbicides

To establish simultaneous C-ELISA, each monoclonal antibody has to selectively react with the corresponding herbicide. Therefore, the cross-reactivities of the monoclonal antibodies with the three herbicides were examined on indirect C-ELISAs. As shown in Table I, each monoclonal antibody was highly specific to each corresponding herbicide, although bensulfuron-methyl and pyrazosulfuron-ethyl are structurally related as a sulfonylurea herbicide. The result suggested that the combination of their monoclonal antibodies would be useful for simultaneous determination of the three herbicides.

Table I. Cross-Reactivities of Monoclonal Antibodies with the Related Herbicides on Indirect C-ELISA

Herbicide	Structure	Cross- reactivity (%)*		
		OC1/ 519-4	OC2/ 14-3	TBC 7-2
Bensulfuron-methyl		100	<0.05	<0.04
Pyrazosulfuron-ethyl		<0.02	100	<0.04
Thiobencarb		<0.01	<0.02	100
Mefenacet		<0.01	<0.02	0.1
Molinate		<0.01	<0.02	<0.01

*The cross-reactivity (%) = 100(A / B).
A shows I_{50} of the corresponding herbicide to each monoclonal antibody.
B shows I_{50} of each herbicide examined.

During rice planting season, many different kinds of herbicides are applied to the paddy fields. Water samples from paddy fields in this season generally contain multiple herbicides. In particular, mefenacet and sulfonylurea herbicides frequently exist in the same water samples because they are often common ingredients of the herbicide mixtures used on rice. Therefore, the cross-reactivities of the monoclonal antibodies with mefenacet were also examined. As shown in Table I, none of the monoclonal antibodies had significant cross-reactivity with mefenacet except for TBC7-2; cross-reactivity being only 0.1%. On the other hand, TBC7-2 was reactive to thiobencarb, but not to molinate, even though both compounds are structurally similar and belong to a category of carbamate herbicides. These results indicate that the reactivity of the monoclonal antibody for each herbicide is little affected by the other non-targeted herbicides in water samples.

Establishment of Simultaneous C-ELISA

Simultaneous C-ELISA can be achieved by using an antibody which cross-react with several pesticides, or by using a series of antibodies which each specifically reacts with their corresponding pesticide. In particular, the latter type of the C-ELISA combined with use of specific antibodies may enable simultaneous and quantitative determination of each pesticide by simply pre-mixing the water sample with the HRP-haptens.

Based on this concept, the simultaneous C-ELISA with OC-1/519-4, OC-2/14-3 and TBC7-2 was developed for determination of bensulfuron-methyl, pyrazosulfuron-ethyl and thiobencarb, as the flow chart shows in Figure 2. Determination of herbicide at the range from 0.20 to 1.5 ng/ml for bensulfuron-methyl, from 0.20 to 4.5 ng/ml for pyrazosulfuron-ethyl and from 0.20 to 3.5 ng/ml for thiobencarb was achieved, as shown in Figure 4. The ranges were defined, such that, the lower limit was 20% of the inhibition and the upper limit was 80% inhibition. These results reveal that the working ranges are sufficiently sensitive because the guidelines for the residual herbicides in drinking water in Japan are 400 ng/ml for bensulfuron-methyl, 100 ng/ml for pyrazosulfuron-ethyl and 20 ng/ml for thiobencarb.

Recovery Study: Determination of the Herbicides Added into Water

Many different kinds of compounds which exist in the water from rice paddies, rivers and ponds etc. may affect the binding reactions of the simultaneous C-ELISA. Therefore, determination of the three herbicides added in the water

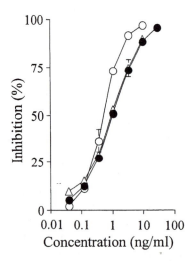

Figure 4.   Reactivities of the monoclonal antibodies with corresponding herbicides on simultaneous C-ELISA.
The reactivity of OC-1/519-4 with bensulfuron-methyl (○), OC-2/ 14-3 with pyrazosulfuron-ethyl (●) and TBC7-2 with thiobencarb (△) are shown. Each value indicates means ± SD (n=6).

samples was examined by the simultaneous C-ELISA. The water samples collected from a pond (Tega-numa, Chiba prefecture, Japan) and a rice paddy were added to each herbicide solution and then applied to the assay. Each of sample assays was carried out 2 times, and the absorbance data obtained were transformed to the herbicide concentrations by the comparison with standard curves. As shown in Table II, the three targeted herbicides were analyzed by the C-ELISA without any significant interference.

These results suggest that the simultaneous C-ELISA is useful for the determination of herbicide residues in water collected from the environment.

Table II. Recovery Study

Herbicide	Added conc.	Recovery (%)	
	(ng/ml)	Paddy water	Pond water
Bensulfuron-methyl	0.37	90	98
	1.1	105	95
Pyrazosulfuron-ethyl	0.37	93	103
	1.1	100	96
Thiobencarb	0.37	104	108
	1.1	101	94

Each value indicates the mean of 2 independent assay results.

Each assay was carried out at triplicate.

Correlation with Instrumental Analyses

To confirm the reliability of the analytical values obtained by simultaneous C-ELISA, these values were compared to those obtained by LC/MS/MS analysis for bensulfuron-methyl and pyrazosulfuron-ethyl and by GC analysis for thiobencarb, respectively.

Surface water samples were periodically collected from the rice paddy where the herbicides were applied, and they were subjected to analysis of the herbicide residues. As shown in Figure 5-1 and -2, the slopes of their equations indicated slightly high bias for C-ELISA relative to LC/MS/MS for bensulfuron-methyl ($y = 0.87x - 0.21$) and a slightly low bias for C-ELISA relative to the LC/MS/MS for pyrazosulfuron-ethyl ($y = 1.08x - 0.40$). Correlation coefficients for bensulfuron-methyl and pyrazosulfuron-ethyl were almost identical ($r = 0.98$ for bensulfuron-methyl and $r = 0.99$ for pyrazosulfuron-ethyl). The slope of the equation for thiobencarb showed a slight high bias for C-ELISA relative to GC ($y = 0.89x - 25.2$) (Figure 5-3), but the correlation coefficient of results for the C-ELISA and the GC was almost identical ($r = 0.96$). The intercept on y-axis was away from the origin compared to those for the other herbicides. This may result

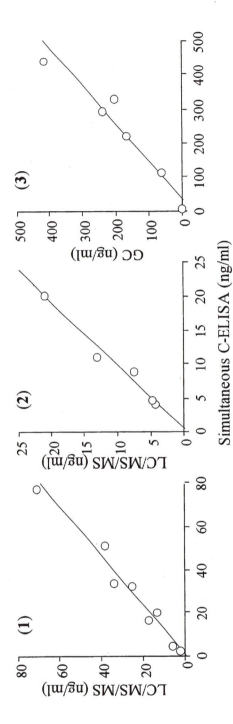

Figure 5. Correlation of analyzed value between simultaneous C-ELISA and instrumental analysis. The correlations for: 1) bensulfuron-methyl; y = 0.87x-0.21, 2) pyrazosulfuron-ethyl; y = 1.08x- 0.40, and 3) thiobencarb; y = 0.89x-25.2 are shown.

from the fact that the determined values for thiobencarb residues are nearly one order of magnitude higher compared to those of the other residues.

Thus, the correlation coefficient for the three herbicides was identical, although slight differences in the biases of the slopes were observed. These results strongly suggest that the herbicide residues in water can be determined by the simultaneous C-ELISA as well as the instrumental analyses.

Comparison with other simultaneous C-ELISA methods

As another simultaneous C-ELISA method which can separately determine plural herbicides in one sample, the antigen coated type indirect C-ELISA was developed previously (4), for the simultaneous determination of 9 kinds of sulfonylurea herbicides. For this C-ELISA, a conjugate of hapten-ovalbumin (multiple antigen) was coated on the wells of 96-well microtiter plate. A sample was added to the wells after the sample was separately mixed with each of 9 kinds of specific antibodies to the target herbicide. This indirect C-ELISA could simultaneously determine each of their herbicides in one sample.

On the other hand, the simultaneous direct C-ELISA we developed could determine each of the herbicides in one sample by mixing with a mixture of three HRP-haptens and then adding to each corresponding well of 96-well microtiter plate. It would be easier to handle than the indirect C-ELISA, because the mixing step is only one time per a sample.

The multiple antigen based C-ELISA could determine structurally related herbicides, but our type of simultaneous C-ELISA could determine herbicides which the structures are different, not only structurally related herbicides. In principle, our C-ELISA would be able to combine the various direct C-ELISAs developed for determination of the other herbicides when their C-ELISAs do not cross-react with bensulfuron-methyl, pyrazosulfuron-ethyl and thiobencarb.

Conclusion

Highly reactive monoclonal antibodies for bensulfuron-methyl, pyrazosulfuron-ethyl and thiobencarb were prepared. The monoclonal antibodies prepared were highly specific to the corresponding herbicides. By using the monoclonal antibodies, the simultaneous competitive enzyme-linked immunosorbent assay (C-ELISA) was achieved. The C-ELISA provided the tool for determination of residues of the three herbicides in water without any pre-treatment of the water.

The simultaneous C-ELISA provides a number of advantages for analysis of water samples, as listed in Table III. Actually, up to 100 or more water samples containing herbicides can be directly determined by the C-ELISA in only 2 to 3 hours.

The water samples from rice paddies often contain several herbicides such as bensulfuron-methyl, pyrazosulfuron-ethyl and thiobencarb. Therefore, the simultaneous C-ELISA would be very useful for simple determination of the herbicide residues.

Table III. Advantage of the Simultaneous C-ELISA for Water Samples

* Very small sample size (only, 1ml)
* Direct determination without any pre-treatment
* Rapid analysis of many samples at once
* Simultaneous determination of plural herbicides in one water sample

Acknowledgment

We express our gratitude to Dr. Hiroshi Kita and Dr. Noriharu Ken Umetsu for the scientific advice and to Dr. Hideyuki Tanaka for the informative advice.

References

1. Jung, F.; Gee, S. J.; Harrison, R. O.; Goodrow, M. H.; Karu, A. E.; Braun, A. L.; Li, Q. X.; Hammock, B. D. *Pestic. Sci.* **1989**, *26*, 303-317.
2. Aga, D. S.; Thurman E. M. *Immunochemical Technology for Environmental Applications*; Aga, D. S.; Thurman E. M., Eds.; ACS Symposium Series 657; American Chemical Society: Washington, DC, 1997; p 1-20.
3. Meulenberg, E. P.; Mulder, W. H.; Stoks, P. G. *Environ. Sci. Technol.* **1995**, *29*, 553-561.
4. Strahan, J.; *Environmental Immunochemical Methods*; Van Emon, J. M.; Gerlach, C. L.; Johnson, J. C., Eds.; ACS Symposium Series 646; American Chemical Society: Washington, DC, 1996; p 65-73.
5. Miyake, S.; Morimune, K.; Yamaguchi, Y.; Ohde, K.; Kawata, M.; Takewaki, S.;Yuasa, Y. *J. Pesticide Sci.* **2000**, *25*, 10-17.
6. Miyake, S.; Ito, S.; Yamaguchi, Y.; Beppu, Y.; Takewaki, S.;Yuasa, Y. *Analytica Chimica Acta* **1998**, *376*, 97-101.

Chapter 8

Translocation and Metabolism of Pesticides in Rice Plants after Nursery Box Application

Shin Kurogochi

Environment and Safety Research, Yuki Research Center, Nihon Bayer Agrochem, K.K., Yuki, Ibaraki 307–0001, Japan

This paper compares the systemic properties and long lasting efficacy of imidacloprid and carpropamid in rice after nursery box application. The physicochemical properties and metabolic stability in both soil and plant influenced the distribution manner of the compounds in rice plants. Metabolic profile indicated that the parent compounds were one of the major components in rice, even at the harvest time, showing the long lasting efficacy of the pesticides. The rather low total radioactive residues found in the rice grains for both compounds, <0.02 mg/kg, stressed the safety for edible parts.

Machine transplanting of rice seedlings has become the dominant method in East Asian countries such as Japan, Korea and Taiwan. The rice seedlings are grown in a nursery box before the machine transplanting. The application of pesticides into the nursery box provides possibilities for labor savings and environment-friendly agricultural practices. One nursery box (size: 0.18 m^2) can supply seedlings which can cover 50 m^2 of a paddy field. Thus the spray area in the box is only 1/280 of whole field application. The premixed formulations, such as insecticide plus fungicide offer also labor savings. The preferred properties of pesticides for the nursery box application should be systemic, long lasting efficacy and safety for humans and the environment. The compounds with the preferred properties can reduce the application frequencies by skipping some of the conventional sprays. The nursery box application has the advantage of directly supplying the active ingredients to the root zone and this enhances the availability of the active ingredients in the field conditions. For example, carpropamid can control rice blast with 1/5 amount of the pesticide in the nursery box application compared with submerged application after transplanting in the paddy field (*1*). This method can also reduce the total amounts of the chemicals applied into the paddy fields per season. These practices lead to the environment-friendly agriculture.

Several insecticides and fungicides have been developed and some are being developed for the nursery box application (Fig. 1 and 2). The nursery box application in Japan has increased significantly and it was announced that 55% of the total cultivation area was treated with nursery box application in 1999. It is estimated that in the near future the ratio of nursery box application will increase to 60 - 70% of total cultivation area of rice.

Among the compounds shown in Fig. 1 and 2, imidacloprid (2,3) and carpropamid (4) are representatives of widely used insecticides and fungicides which have been recently developed as rice pesticides for nursery box application as the main use pattern. The water solubility of imidacloprid and carpropamid are 610 mg/L and 4 mg/L (20 °C), respectively. The differences in the water solubility might affect the translocation behavior in the rice plants. The ^{14}C-radiolabelled compounds were used to conduct the translocation and metabolism studies.

Imidacloprid

Imidacloprid has been launched as the first neonicotinoide insecticide since 1992 in Japan. Imidacloprid has been used as an effective insecticide against sucking insects (green rice leafhopper and various rice planthoppers), and other pests of rice (rice water weevil, rice thrips, rice leaf beetle and rice leafminer) with nursery box application. Imidacloprid is moderately water soluble and is characterized as a systemic insecticide.

Translocation of imidacloprid in rice with hydroponic culture

The translocation of imidacloprid in rice plants was investigated with methylene-^{14}C-radiolabelled compound using hydroponic culture system. The distribution of the radioactivity in the rice seedlings was determined by autoradiography. Fig. 3 shows the autoradiograms at 1 and 8 days after treatment (DAT). Imidacloprid was translocated acropetally and accumulated near the tip of the leaves. There was no apparent accumulation of radioactivity observed in the roots and lower parts of the stem. The sucking insects such as green leafhopper suck at first in the xylem and then phloem. The imidacloprid transported in xylem with the water stream is available for a direct attack to such sucking insects (5). A phloem-sucking insect, such as aphid, is possibly affected with imidacloprid transported into phloem owing to its water soluble characteristics (6).

Metabolism of imidacloprid in rice

The application of the ^{14}C-radiolabelled compound with modified nursery box method shown in Fig. 4 was chosen due to the technical difficulty of tracing the practical application of granular formulation to seedlings in the nursery box. The modified method was justified with the quantitative application of the ^{14}C-compound in the root zone of rice and the limited duration in the nursery box after application of pesticide comparing to the whole growing period. Recommended use pattern is to apply formulation to the nursery box at the timing from 0 to 3 days before transplanting. Methylene ^{14}C-imidacloprid, 5.58 MBq/mg, 1.6 mg/pot (0.32 kg a.i./ha), was applied with 1 mL of acetone solution. Three rice seedlings (*Oryza sativa* L. cv. Koshihikari) were planted in a pot with 500 cm^2 surface area and the rice plants were grown in a radioisotope-green house with the temperature of 16-30 °C and the relative humidity of 36-79%. Forage was sampled at 65 DAT and straw and grain were sampled at 124 DAT. The radioactive residues in the plant tissues were separated, quantified and identified with analytical instruments such as HPLC, TLC linear analyzer, mass spectrometer and NMR.

The total radioactive residues (TRR) expressed as parent compound equivalent (mg/kg) in the forage, straw and grain were 0.378, 1.314 and 0.014 mg/kg, respectively. Metabolic profile of imidacloprid in rice plants is shown in Fig. 5. The major metabolite found in the forage and the straw was the guanidine compound (refer the structure in Fig. 6). It was however not a major metabolite in the grain. Imidacloprid was one of the major residues in all tissues, it was a major component identified also in the grain in percentage-wise, but the absolute residue level was significantly low (<0.002 mg/kg). Bound residues especially in the grains accounted for more than 2/3 of the TRR in the grain.

The absolute total residue amount in the grain was too low for analysis. Therefore, the further investigation of the bound residues in the grain was conducted with the samples harvested with 4 fold excess application of ^{14}C-imidacloprid. The bound residues in the grain were further fractionated into natural constituent fractions including starch fraction (7). The crude starch fraction (5 g) was hydrolyzed with 300 mL of 0.5 N hydrochloric acid by refluxing for 6 h. The solution was neutralized with sodium hydroxide solution and washed with 300 mL of ethyl acetate. The aqueous solution was concentrated to 20 mL and then 20 mL of ethanol was added to precipitate

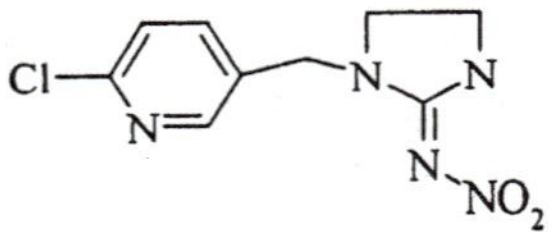

carbosulfan

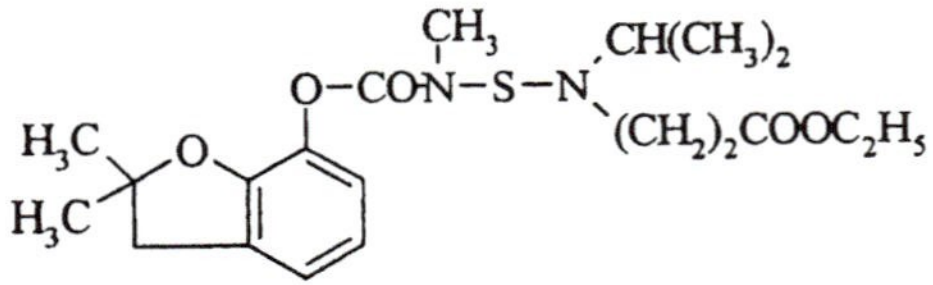

benfuracarb

*tefranitodine

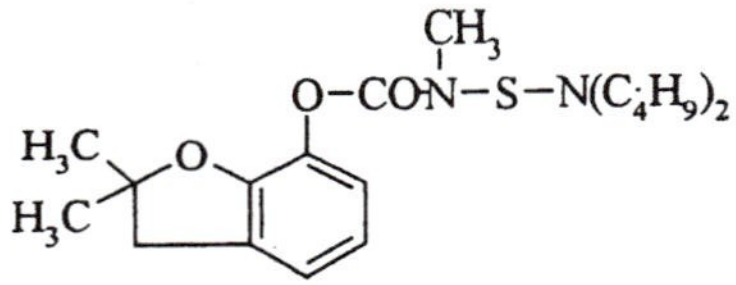

fipronil

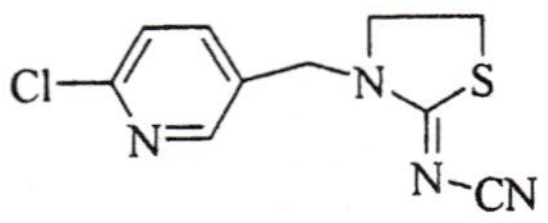

imidacloprid

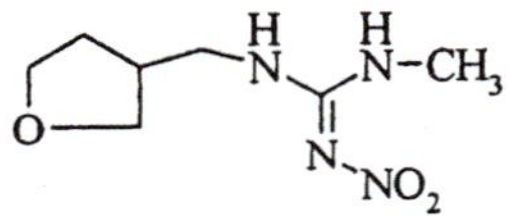

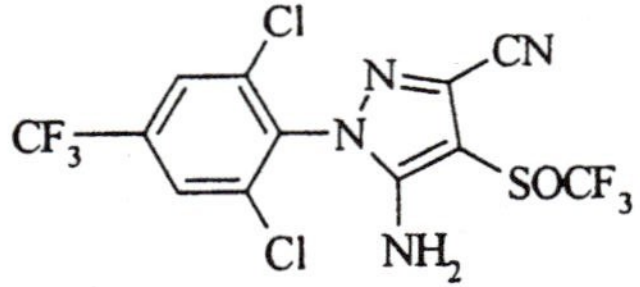

*thiacloprid

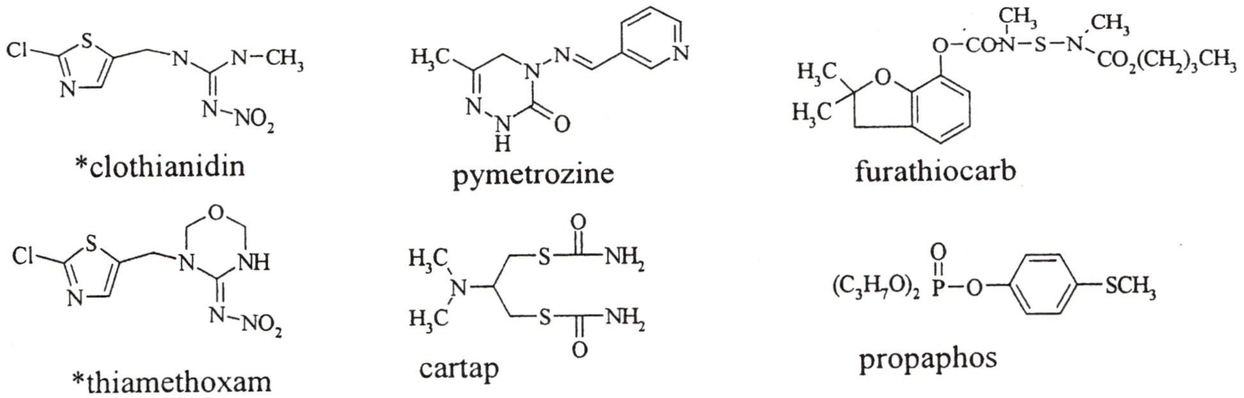

Figure 1. Insecticides developed or *developing for nursery box application in Japan.

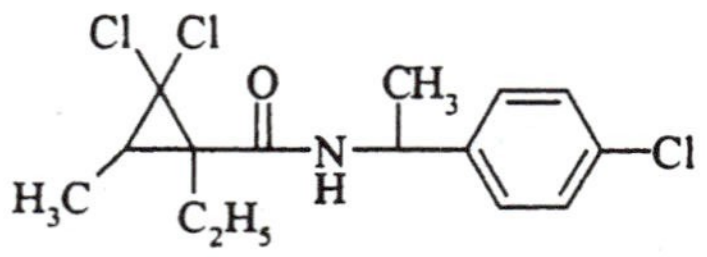

azoxystrobin

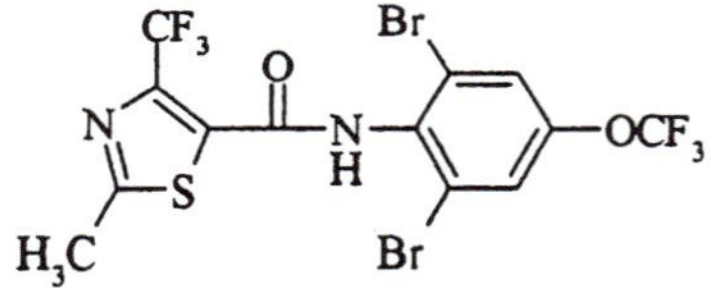

thifluzamide

acibenzolar-S-methyl

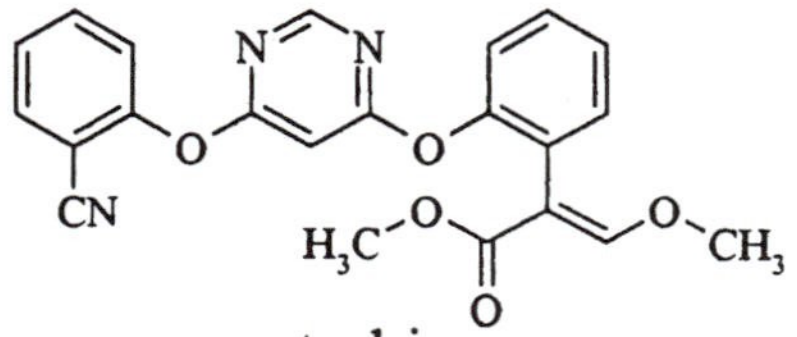

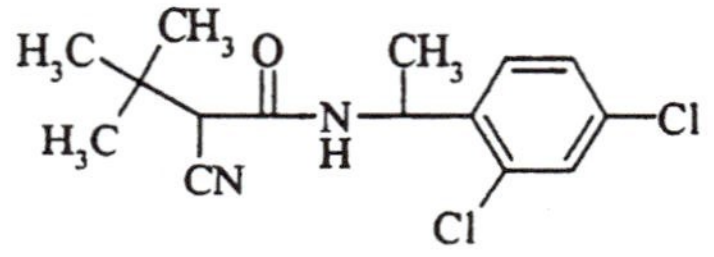

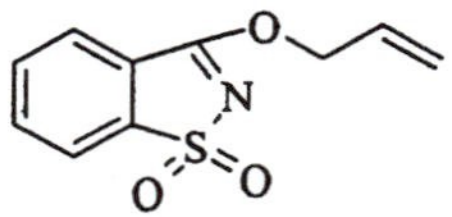

probenazole

carpropamid

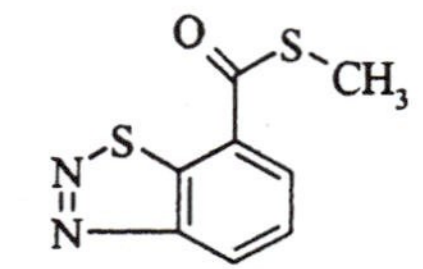

*diclocymet

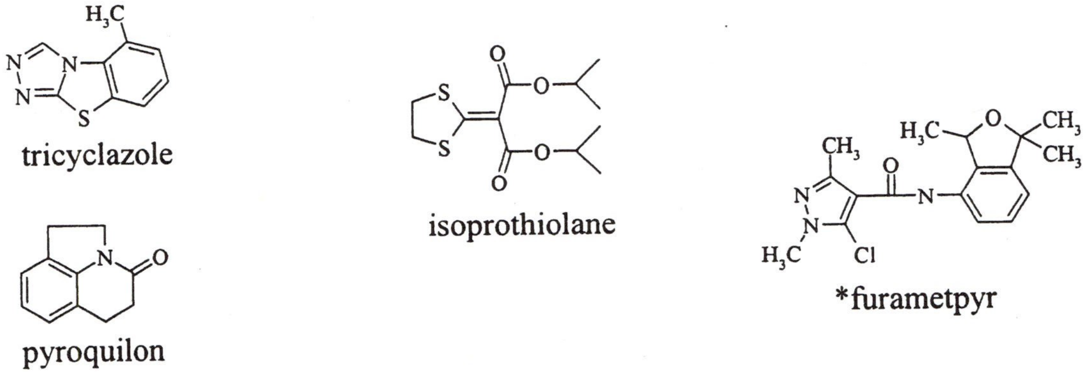

*Figure 2. Fungicides developed or *developing for nursery box application in Japan.*

146

[Methylene-¹⁴C] imidacloprid   [Phenyl-UL-¹⁴C] carpropamid

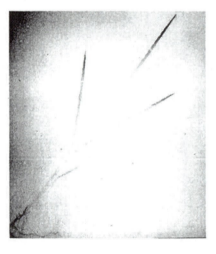 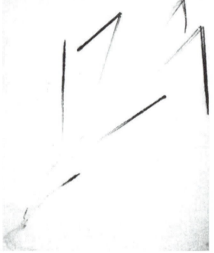

Figure 3. Autoradiograms of rice seedlings, which were exposed to the hydroponic solution including ¹⁴C-imidacloprid, left 1DAT and right 8 DAT

1. Remove surface
   water
2. Insert glass
   cylinder
3. Take out soil
4. Refil semi-dry soil
5. ^{14}C-compound in
   organic solvent
6. Stand for 1 h

7. Transplant
   3 seedlings
8. Cover with soil

9. Irrigate

10. Remove glass
    cylinder after
    1 week

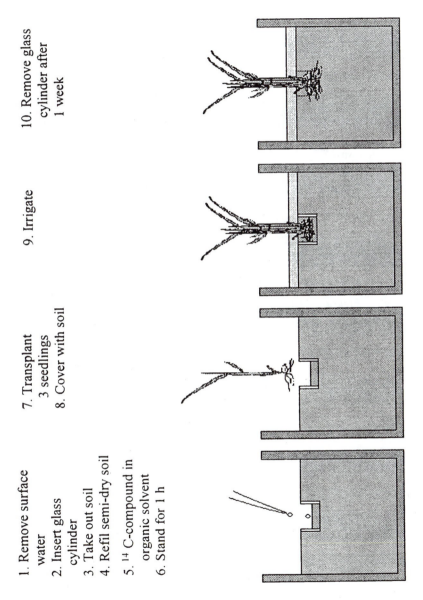

Figure 4. Modified nursery box application of ^{14}C-pesticides

148

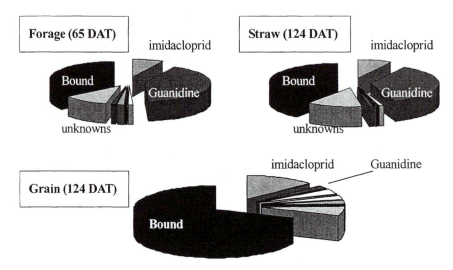

Figure 5. Metabolic profile of imidacloprid in rice plants

**Table I. Incorporation of ^{14}C into starch in the grain after complete
degradation of imidacloprid to CO_2**

Material	Total (Bq)	Dry weight (g)	Specific activity Bq/g
Grain	24,167	67.88	356
Crude starch fr.	1,410	5.02	281
Penta-acetylglucose			
1st Crystal	809	6.57	123
2nd Crystal	710	5.79	123
3rd Crystal	635	5.02	126
		Ave.	124

sodium chloride. The supernatant was concentrated to dryness and then acetylated with 1.5 g of sodium acetate and 60 mL of acetic anhydride at 90-100 °C for 3 h. The penta-acetylglucose was extracted with organic solvent and the solution was washed and dried over anhydrous sodium sulfate and concentrated to give crude penta-acetylglucose. The acetylglucose was recrystallized from 1) *n*-hexane/diethyl ether, 2) and 3) from ethanol. The specific radioactivity of grain, starch and penta-acetylglucose are shown in Table I.

In order to know how much percentage of TRR in the grain is attributed to the natural constituent "starch", the following equation was used:

$$\% \text{ of starch in TRR} = \frac{\text{Specific radioactivity of starch (A Bq/g) x 0.7}}{\text{Specific radioactivity of grain (B Bq/g)}} \times 100$$

$$\text{Specific radioactivity of starch (A Bq/g)} = \frac{(C_{16}H_{22}O_{11})_n}{(C_6H_{10}O_5)_n} \times 124$$

0.7: Starch content in rice grain

The calculation showed 58.7% of TRR in the grain was attributed to natural constituent "starch". The results revealed that imidacloprid was obviously decomposed to $^{14}CO_2$ in the paddy soil (8). The released $^{14}CO_2$ were then taken up by the plants for the regular photosynthetic pathways. The photosynthetic products (i.e. glucose) are finally transported to the rice grains and incorporated into the natural constituent starch.

The proposed metabolic pathways are shown in Fig. 6. The main metabolite was guanidine in the forage and straw, it further converted to urea, 6-chloronicotinic acid (6-CNA) or bound residues. Other metabolites were hydroxylated at imidazolidine ring and followed by dehydration to convert to the olefin. All metabolites found in the rice plants were also identified in the animal metabolism study (9).

Carpropamid

Carpropamid is a new melanin biosynthesis inhibitor that has new action site against dehydratases in the rice blast fungus and has no direct fungicidal activity *(10)*. Previous melanin biosynthesis inhibitors, such as tricyclazole and pyroquilon (Fig. 2) were much water soluble and inhibit reductases in the rice blast fungus. Carpropamid is less water soluble and would be expected to translocate differently in rice plants and to have a different mode of protection. Carpropamid has been launched since 1997 in Japan as nursery box application. Its formulations are available as single or in combination with other pesticides such as imidacloprid.

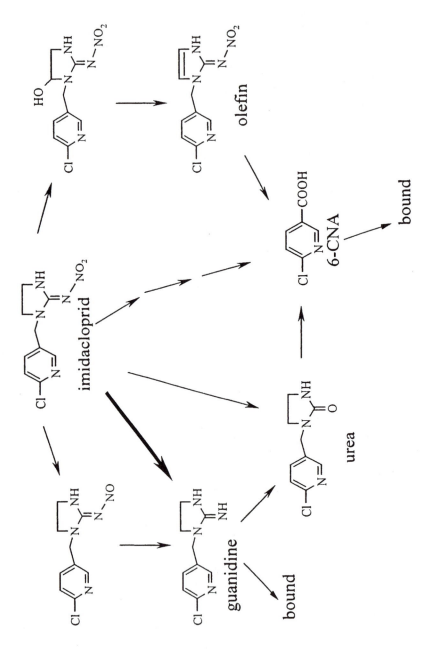

Figure 6. Proposed metabolic pathways of imidacloprid in rice plants

Translocation of carpropamid in rice with hydroponic culture

The translocation of carpropamid in rice plant was investigated with phenyl-UL-^{14}C-radiolabelled compound using hydroponic culture (*11*). The rice seedlings were measured by radioluminography. Fig. 7 shows the results at 1 DAT. Carpropamid was translocated acropetally and uniformly distributed in the whole rice plant. A significant accumulation was observed in the roots and lower parts of the rice stem. The uniform distribution could be important for the protection of the rice plant from the invasion of rice blast fungus at any part of the rice plant. The distribution to the plant surfaces was also examined by rinsing with organic solvent (*12*). It indicated that ca. 20-40% of the carpropamid in the shoot is localized on the surface of the plant. This part should have a direct contact with the mycelium or appressorium of the fungus and then inhibit the melanin biosynthesis.

Figure 7. Autoradiograms of rice seedlings which were exposed to the hydroponic solution including ^{14}C-carpropamid at 1DAT showing photograph (left) and Autoradiogram (right)

When autoradiograms were observed in detail (Fig. 8), the location of radioactivity was observed in more or less regular spots along with the vascular bundles. The radioactivity in the spot was not rinsed away with organic solvents suggesting that the active ingredient and/or its metabolites concentrated/accumulated in the vascular bundles. Those spots were also observed in the roots and lower parts of shoots. These spots are expected to be

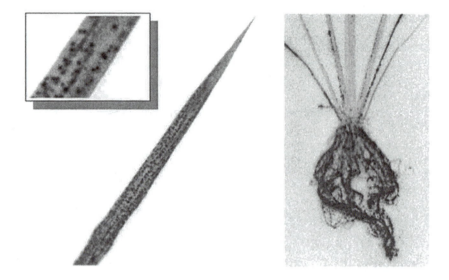

Figure 8. Translocation of ^{14}C-carpropamid in rice plants at 7 DAT with hydroponic culture

utilized for the stable and long lasting supply of active ingredient to newly developing shoots and leaves.

Metabolism of carpropamid in rice

The application of ^{14}C-compound with modified nursery box method shown in Fig. 3 was chosen as mentioned in the imidacloprid section. Phenyl-UL-^{14}C-carpropamid, 6.73 MBq/mg, 2.0 mg/pot (0.4 kg a.i./ha), was applied with 1 mL of acetone solution. Three rice seedlings (*Oryza sativa* L. cv. Koshihikari) were planted in a pot with 500 cm^2 surface area and the rice plants were grown in a radioisotope-green house under the similar conditions as of imidacloprid. Forage was sampled at 67 DAT, and straw and grain were sampled at 115 DAT. The radioactive residues in the plant tissues were separated, quantified and identified with the methods used in the imidacloprid metabolism studies and new technology such as bioimaging analyzer and LC-MS/MS.

The total radioactive residues expressed as parent compound equivalent (mg/kg) in the forage, straw and grain were 0.556, 1.632 and 0.012 mg/kg, respectively. The metabolic profile of carpropamid in rice plants is shown in Fig. 9. The major component found in the forage, the straw and the grain was carpropamid itself. The absolute residue level of carpropamid in the grain was significantly low (<0.01 mg/kg).

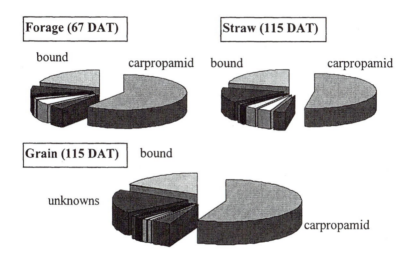

Figure 9. Metabolic profile of carpropamid in rice plants

The proposed metabolic pathways are shown in Fig. 10. The main metabolic reaction was oxidation at the free methyl group of the cyclopropane ring, forming the alcohol. It was subsequently oxidized to the carboxylic acid or conjugated with fatty acids, forming the alcohol esters, or with glucose, forming the alcohol glucoside. Oxidation of the active ingredient in the phenyl ring, leading to the phenol, occurred to only a fairly small extent. The amide bond between the two parts of the molecule is evidently stable, as no cleavage products were detected. These metabolic pathways were also detected in the animal metabolism studies (*11*).

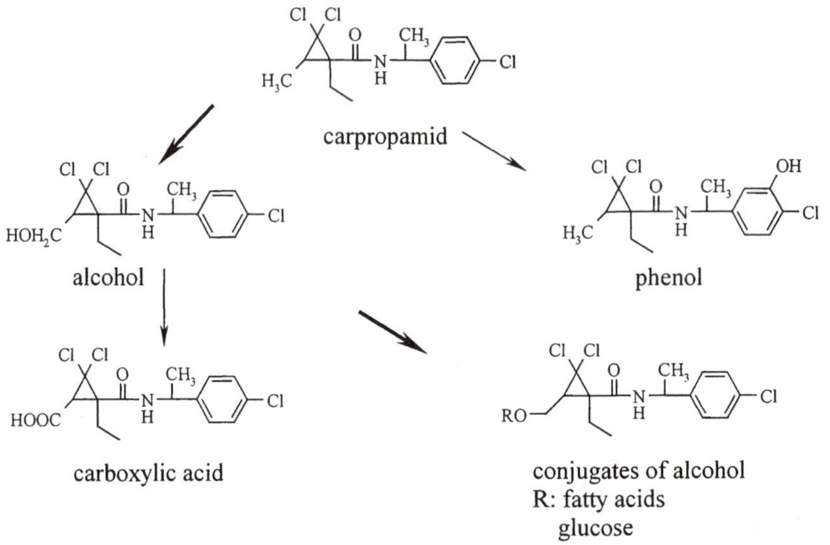

154

Figure 10. Proposed metabolic pathways of carpropamid in rice plants

Comparison of imidacloprid and carpropamid on translocation and metabolism in rice

Although both imidacloprid and carpropamid showed systemic properties in rice plants, the translocation modes were quite contrasting. Moderately water soluble imidacloprid passed through the root, stem and then dominantly accumulated in the tips of the leaves. Less water soluble carpropamid showed also acropetal translocation but with uniform distribution. Carpropamid was the dominant residue generally in rice plants. Because it is easily recrystallized in water, the radioactive spots observed in the autoradiograms should be recrystallized carpropamid in the plants. This phenomenon probably allows the uniform distribution to the whole plant also the supply of carpropamid to the surface of the plant. Therefore, an effective protection on the plant surface is possible.

Both parent compounds were found at least to be one of the major residues in the rice plants even at the harvest time. This supports the long lasting efficacy of both compounds. Although imidacloprid is more quickly metabolized than carpropamid, the biological efficacy was expected for a relatively long period.

As long as the supply of imidacloprid from the root zone is sustained, the control of sucking insects is possible. In the control of the brown rice leafhopper by the nursery box application, the biological control could continue up to 60 days after transplanting (*13,14*). Carpropamid could be stabilized by recrystallization in rice plant, because of the protection from the metabolism enzymes in rice. This phenomenon probably makes the long lasting efficacy possible. Under moderate infection pressure of rice blast diseases, a nursery box application can control leaf blast and panicle blast through a whole season of rice cultivation period (*15*).

In general, nursery box application affords a long pre-harvest interval and thus lower residues in the grain is reasonable. Both compounds were found with only trace levels in the grain (<0.02 mg/kg as TRR). This indicates nursery box application gives the quite low levels of residue in the edible grain.

From the data of two different compounds, imidacloprid and carpropamid, they proved to be excellent compounds for rice nursery box application from the view of labor savings and safety for humans and the environment.

Acknowledgement

The author thanks his colleagues who were involved in the metabolism studies conducted in Yuki Research Center in NBA. For imidacloprid: Ms. Maruyama, M.; Mr. Araki, Y.; Mr. Sakamoto, H.; Ms. Muraoka, C.; Mr. Hoshino, T. For carpropamid: Dr. Koester, J.; Ms. Muraoka, C.; Mr. Sato, M.; Dr. Ishikawa, K.; Mr. Morishima, N. He also appreciates Mr. Yasui, K., Mr. Otsu, Y., Mr. Itoh, S. and Dr. Willms, H. in NBA for their technical information and support. Dr. Koester (Bayer AG) kindly proofed the manuscript. He thanks Dr. Fritz, R. (Bayer AG) and Dr. Ueyama, I. (NBA) for the support and approval of this publication.

References

1. Kurahashi, Y. *Noyaku Kenkyu* **1996**, 42 (No.168), 60-65
2. Shiokawa, K.; Tsuboi, S.; Kagabu, S.; Moriya, K. JPN Patent 1,807,569 1993
3. Shiokawa, K.; Tsuboi, S.; Iwaya, K.; Moriya, K. *J. Pesticide Sci.* **1994**, 19, 329-332

156

4. Kurahashi, Y.; Kurogochi, S.; Matsumoto, N.; Kagabu, S. *J. Pesticide Sci.* **1999**, 24, 204-216

5. Maruyama, M.; Obinata, T. *Noyaku Kenkyu* **1995**, 42 (No.165), 19-26

6. Troeltzsch, C.-M.; Fuehr, F.; Wieneke, J; Elbert, A. *Pflanzenschutz-Nachrichten Bayer* **1994**, 47, 241-287

7. Rauchaud, J.; Moons, C.; Meyer, J. A. *Pestic. Sci.* **1979**, 10, 509-518

8. Klein, O. 8th Int. Cong. Pestic. Chem., Washington, Poster Ses. 2B (No. 157), 1994

9. Klein, O. 8th Int. Cong. Pestic. Chem., Washington, Poster Ses. 2B (No. 367), 1994

10. Kurahashi, Y.; Hattori, T.; Kagabu, S.; Pontzen, R. *Pestic. Sci.* **1996**, 47, 199-202

11. Kurogochi, S.; Koester, J. *Pflanzenschutz-Nachrichten Bayer* **1998**, 51, 219-244

12. Kurogochi, S. *Noyaku Kenkyu* **1996**, 43 (No.170), 32-41

13. Ishii, Y.; Kobori, I.; Araki, Y.; Kurogochi, S.; Iwaya, K.; Kagabu, S. *J. Agric. Food Chem.* **1994**, 42, 2917-2921

14. Iwaya, K.; Maruyama, M.; Nakanishi, H.; Kurogochi, S. *J. Pesticide Sci.* **1998**, 23, 419-421

15. Kurahashi, Y.; Sakawa, S.; Kinbara, T.; Tanaka, K.; Kagabu, S. *J. Pesticide Sci.* **1997**, 22, 108-112

Chapter 9

The Use of Native Prairie Grasses to Degrade Atrazine and Metolachlor in Soil

Shaohan Zhao[1], Ellen L. Arthur[2], and Joel R. Coats[3]

[1]ABC Laboratories, 7200 East ABC Lane, Columbia, MO 65202
[2]Bayer Corporation, 17745 South Metcalf, ACR II, Stilwell, KS 66085
[3]Department of Entomology, Iowa State University, 116 Insectary,
Ames, IA 50011–0001

The ability of native prairie grasses, big bluestem (*Andropogon gerardii* Vitman), Yellow indiangrass (*Sorghastrum nutans* L.), and switchgrass (*Panicum virgatum* L.), to degrade atrazine and metolachlor was evaluated in two soils denoted as Alpha and Bravo soils. Vegetation significantly decreased the amount of remaining atrazine in Alpha soil when the concentration of atrazine before vegetation was 93 μg g^{-1}, but had no effect on the degradation of atrazine when it was 4.9 μg g^{-1}. The significant effect of the plants on atrazine degradation in Alpha soil occurred at 57 days after the transplanting of vegetation, but not at 28 days after the transplanting of vegetation. The grasses did not enhance the degradation of atrazine in Bravo soil due to the population of atrazine-degrading microorganisms in that soil. The native prairie grasses had a significant positive effect on the enhanced degradation of metolachlor in both soils, and the significant effect was observed at 28 and 57 days after the transplanting of vegetation in Alpha and Bravo soil, respectively. NH_4NO_3 had no effect on the degradation of atrazine and metolachlor in either soil. Our results indicate that it is feasible to use the native prairie grasses to help remediate the soils contaminated with high concentrations of atrazine and metolachlor, especially in the absence of the indigenous atrazine or metolachlor degraders.

Introduction

One problem associated with the widespread use of atrazine (ATR, 2-chloro-4-ethylamino-6-isopropylamino-s-triazine) and metolachlor (MET, 2-chloro-N- (2-ethyl-6-methylphenyl)-N- (2-methoxy-1-methylethyl) acetamide) as preemergent herbicides to control grass weeds and broadleaf weeds is the contamination of surface and ground waters (1-3). ATR and MET have been frequently detected in rivers and streams of the Midwestern United States (4-6). ATR occasionally has exceeded the maximum contamination level for ATR in drinking water (3 μg L^{-1}) set by the U. S. Environmental Protection Agency (3, 6). Effective and low-cost remediation approaches at the sites contaminated with ATR and MET are needed.

There is current interest in the use of plants to remediate contaminated soils, sediments, and water, termed phytoremediation. Phytoremediation is an aesthetically pleasing and cost-effective approach for the treatment of hazardous waste sites (7). Plants may act directly on organic compounds via uptake of organics and biotransformation of the organics to less toxic metabolites, and/or indirectly degrade them via the rhizosphere effect (8, 9). The uptake is influenced by the physicochemical properties of the compounds, the characteristics of the plant species, and environmental conditions (9-14). Plants can take up moderately hydrophobic organics (octanol-water partition coefficients, Log K_{ow} = 0.5-3) quite effectively (12). More lipophilic compounds (Log K_{ow} > 3.0) can better partition into roots, but they can not easily be transported within the plants. More lipophobic compounds (Log K_{ow} < 0.5) are not sufficiently sorbed to roots or actively transported through plant membranes (12). Plant characteristics, such as root surface area, could substantially alter adsorption of an organic compound to roots. A large portion of the applied ^{14}C-ATR (91%) in soil has been shown to be taken up by poplar cuttings (Populus deltoides nigra DN34) (7). However, only 28, 9.9, and 0.51% of the applied ATR was taken up in corn, Kochia soparia, and barley, respectively (14-16). A high transpiration rate also can increase the uptake of organic compounds (9,13,14). Uptake by plants also depends on the soil concentration of the chemicals and the plant species (9). The ATR uptake in sudangrass [Sorghum sudanense (Piper) Stapf, var. Piper], grain sorghum [Sorghum bicolor (L.) Moench], and corn (Zea mays L.) was positively correlated with the concentration in soil (17). However, higher uptake for the lower soil concentration of ATR was reported in barley (16). The amount of uptake also varies with different soil types used (7,14).

The rhizosphere is the region immediately surrounding the roots of a plant. It serves as an enrichment zone for increased microbial activity. Plants not only release exudates in the rhizosphere for microbial growth or cometabolism, but also harbor microbial consortia and mycorrhizal fungi on root surfaces. A great density and diversity of microorganisms occur in the rhizosphere (*14,18-20*). As a result, enhanced metabolic or cometabolic biotransformation frequently occurs in the rhizosphere. Studies have demonstrated the increased degradation of organics in the rhizosphere of a variety of plant species (*20-24*). However, the rhizosphere effect is short-lived in the absence of plants (*25*), and the relevance of soil-only data to *in situ* plant-microbial interactions is limited. There are also cases in which the rhizosphere has no effect on the mineralization of organic compounds (*14,23*). The objective of this study was to determine the effectiveness of native prairie grasses on the degradation of ATR and MET in two soils.

Materials and Methods

Chemicals. ATR (92.2% pure) and MET (97.3% pure) for treating the soils were obtained from Novartis Crop Protection (Greensboro, NC). ATR (98% pure for analytical standard) was purchased from Chem Service (West Chester, PA).

Soils and Plants. Soils were obtained from two agrochemical dealer sites in Iowa. The two sites, denoted as the Alpha site and Bravo site, are located in northwest Iowa and in central Iowa, respectively. Surface soils (top 15 cm) were randomly collected from the vegetated areas by using hand trowels, and combined for each replication. Three replications of soils were taken, sieved (2.4 mm), and stored in the dark at 4°C until needed. Soils were analyzed by standard methods to determine physical and chemical properties. The Alpha soil had a sandy loam texture with 68% sand, 21% silt, and 11% clay. The organic matter, total nitrogen, pH, and cation exchange capacity were 2.5%, 0.08%, 7.8, and 10.0 meq/100g, respectively. The Bravo soil had a loam texture with 32% sand, 50% silt, and 18% clay. The organic matter, total nitrogen, pH, and cation exchange capacity were 3.9%, 0.22%, 7.5, and 14.1 meq/100g, respectively.

Three species of native prairie grasses, big bluestem (*Andropogon gerardii* Vitman), Yellow indiangrass (*Sorghastrum nutans* L.), and switchgrass (*Panicum virgatum* L.), were utilized in this study. The mixture of grasses was planted in a small tray in the greenhouse until the height range of the grasses was between 10 to 20 cm. Then the root soils of the grasses were washed off with tap water, and the grasses were transplanted into Ray Leach "Cone-Tainers"

(Stuewe & Sons, Inc., Corvallis, Oregon) along with glass wool filling the bottom of each cone and the soils treated with ATR and MET and aged for various days in the following experiments. Each cone contained 6 to 12 grass plants (a mixture of the three species of native prairie grasses).

Alpha Soil Study I. A treating solution consisting of a mixture of ATR and MET was prepared in acetone and applied to Alpha soils at a concentration of 100 μg g^{-1} soil (dry weight) for ATR and 25 μg g^{-1} soil (dry weight) for MET. The chemicals were aged for 13 days in the greenhouse at a temperature of 27 ± 2 °C before the following four treatments were added: addition of NH_4NO_3 (equivalent to 89.7 kg ha^{-1}), vegetation, addition of both NH_4NO_3 (equivalent to 89.7 kg ha^{-1}) and vegetation, and addition of only a phosphate buffer (control). Five mL of water was added per 120-g soil (dry weight) on a weekly basis during the aging period. There were four replications for each treatment, and each replication contained 80 g of soil (dry weight). Analysis by gas chromatography indicated that the soils contained an average of 93 ± 5 μg g^{-1} of ATR and 24 ± 2 μg g^{-1} of MET before the treatment with vegetation and NH_4NO_3. The treated soils were placed in cones, and then the grasses were transplanted in the cones. The cones were kept in the greenhouse, and water was added to the soils on a daily basis to maintain adequate moisture until the end of the study. Concentrations of ATR and MET were determined at 57 days post treatment with vegetation and NH_4NO_3. The reported percentage of remaining ATR or MET at 57 days post vegetation and NH_4NO_3 was calculated by dividing the concentrations of ATR or MET at 57 days post treatment with vegetation and NH_4NO_3 by the concentrations before the treatment with vegetation and NH_4NO_3, then multiplying by 100.

Alpha Soil Study II. Alpha soils were treated uniformly with a mixture of ATR and MET by using acetone as the solvent, providing 100 μg of ATR g^{-1} soil (dry weight) and 25 μg of MET g^{-1} soil (dry weight). After fortification with the chemicals, the soils were placed in cones, and were aged for 50 days in the greenhouse at a temperature of 27 ± 2 °C before the following two treatments were added: vegetation, and control treatment (no vegetation). Five mL of water was added to each cone (103-g soil, dry weight) on a weekly basis during the aging period. There were eight replications for each treatment, and each replication contained 75 g of soil (dry weight). Analysis by gas chromatography indicated that the soils contained an average of 84 ± 23 μg g^{-1} of ATR and 17 ± 3 μg g^{-1} of MET before the treatment with vegetation. The aged soils were placed in cones, and then the grasses were transplanted in the cones. The cones were kept in the greenhouse, and water was added to the soils on a daily basis to maintain adequate moisture until the end of the study. Concentrations of ATR and MET were determined at 28 days post vegetation. The reported percentage of remaining ATR or MET at 28 days post vegetation was calculated by dividing the concentrations of ATR or MET at 28 days post vegetation by the concentrations before vegetation, then multiplying by 100.

Alpha Soil Study III. The Alpha soil used in this study had been fortified with a mixture of ATR and MET (200 μg g^{-1} soil each) and aged for 67 days before remediation, and had been remediated by using the same species of the grasses as in the current study and bacteria for 213 days. At the beginning of the current study, the average concentrations of ATR and MET were 4.9 ± 4.0 and 71 ± 18 μg g^{-1} soil (dry weight), respectively. The soil was divided into two treatment groups. One half of the soil was placed in cones, and the grasses were transplanted in the cones. The other group of cones contained soil only. Each cone contained 90 g soil (wet weight). Each group contained four replicates. The cones were kept in the greenhouse, and water was added to the soils on a daily basis to maintain adequate moisture until the end of the study. Concentrations of ATR and MET were determined at 56 days post vegetation. The reported percentage of remaining ATR or MET at 56 days post vegetation was calculated by dividing the concentrations of ATR or MET at 56 days post vegetation by the concentrations before vegetation, then multiplying by 100.

Bravo Soil Study. The procedures exactly followed the Alpha soil study I except that the Bravo soil was used, and the chemicals were aged for 14 days before the treatment with vegetation and NH_4NO_3. The average concentrations of ATR and MET before the treatment with vegetation and NH_4NO_3 were 69 ± 23 and 31 ± 2 μg g^{-1}, respectively.

Chemical Analysis. After the appropriate time period, the soils were extracted with ethyl acetate three times by mechanical agitation. The extracts were concentrated with a rotary evaporator. The concentrated extracts were analyzed using a Shimadzu GC-9A gas chromatograph (Shimadzu Corporation, Kyoto, Japan) equipped with a flame-thermionic detector. Chromatographic conditions were: injector temperature 250 °C; detector temperature 250 °C; column: glass packed with 80/100 mesh 3% OV-17 (Supleco Inc., Bellefonte, PA), 1.2 m length x 3 mm i.d.; column temperature 230 °C; carrier gas helium; flow rate 45 mL min^{-1}. The extraction efficiency for ATR and MET was 107 ± 9% and 98 ± 0.1%, respectively. The quantitation limit = (the concentration (μg mL^{-1}) of the standards required to give a signal-to-noise ratio of 2:1) * (10 mL of the soil extract)/25 g soil. The quantitation limit for ATR and MET was 0.078, and 0.313 μg g^{-1} soil, respectively. The significance of the differences of the percentage of remaining ATR or MET between vegetated soils and unvegetated soils or between the soils amended with NH_4NO_3 and the soils without NH_4NO_3 was tested by analysis of variance.

Results and Discussion

The mixture of the prairie grasses significantly enhanced the degradation of ATR in the Alpha soil study I (Table I). Our result is consistent with the findings of Arthur et al., who reported that *K. scoparia* significantly decreased the extractable ATR in Alpha soil (*15*). However, the grasses did not significantly enhance the degradation of ATR in the other studies (Table I). The duration of the growth of the grasses in the Alpha soil study I and III was very similar after they were transplanted to the contaminated soils; however, the effect of vegetation on the degradation of ATR was different. The main difference between the study I and III is the different ATR concentrations before vegetation. The average concentration of ATR before vegetation in the study I was much greater than that in the study III. Our results indicate that vegetation can degrade ATR more efficiently when the concentration of ATR is high than when the concentration of ATR is low in soil.

Table I. The influence of a mixture of three native prairie grasses on the degradation of atrazine in Alpha and Bravo soils. Data are reported as percentage of remaining atrazine at the end of each study

	Vegetation	*No Vegetation*	*Pr > F*	*Standard Error of Mean*
Alpha Soil Study I	2.3	21	0.0198	4.92
Alpha Soil Study II	38	42	0.6932	6.45
Alpha Soil Study III	90	85	0.6420	7.01
Bravo Soil Study	14	13	0.4817	0.01

In the Alpha soil study II, the concentration of ATR before vegetation was as high as that in the Alpha soil study I. However, the duration of the growth of the grasses in soils in the study II was only half of that in the study I after they were transplanted to the contaminated soils. The longer duration of the growth of the grasses in the study I probably resulted in the greater uptake of ATR into the plants or greater degradation of ATR in the rhizosphere.

Vegetation had no effect on increasing biodegradation of ATR in Bravo soil. This failure appears to be related to the effective mineralization of ATR by indigenous ATR- degraders in that soil. The number of indigenous ATR-degraders is much higher than that in Alpha soil (*15, 26*). The large indigenous population of ATR-degraders in Bravo soil was effective in mineralizing ATR. Struthers et al. reported that approximately 50% of the applied ^{14}C-ATR (50 µg g[-1]) was mineralized within 63 days after the treatment of ATR in Bravo soil

(*26*). Perkovich et al. noted that 62 and 49% of the applied ^{14}C-ATR (50 μg g[-1]) was evolved as [14]CO$_2$ after 36 days of incubation in *K. scoparia* rhizosphere soil and non-rhizosphere soil from the Bravo site, respectively (*22*).

The addition of the prairie grasses significantly reduced the concentrations of MET in all the studies (Table II). Other researches have showed that corn and aquatic plants, such as coontail (*Ceratophyllum demersum*), American elodea (*Elodea canadensis*), and common duckweed (*Lemna minor*), were effective in enhancing the degradation of MET in soil (*24*) or water (*27*), respectively. In the Alpha soil study II, the grasses significantly decreased the amount of remaining MET, but not ATR. It may be related with the greater water solubility and lipophilicity of MET (530 mg L[-1] at 20 °C and K$_{ow}$ of 2820, respectively) compared with those of ATR (33 mg L[-1] at 27 °C and K$_{ow}$ of 219, respectively). The greater water solubility and the lower concentration of MET before vegetation make it possible that a larger percentage of the applied MET is dissolved in the soil water than for ATR. The concentration of the chemical in soil water is a major factor influencing the direct uptake of the chemical through the plant roots (*9*). As a result, a larger percentage of the applied MET may be taken up by the plants compared with that of ATR in the same time frame. In addition, uptake of MET by the plant roots may be greater than that of ATR due to the higher lipophilicity of MET. In a previous study, a significantly greater amount of ^{14}C was found in the plant tissues of the ^{14}C-MET-treated water systems than in the ^{14}C-ATR-treated water systems when both chemicals were applied to water at same concentration (*27*). The lower water solubility and lipophilicity of ATR may explain why more time is needed for the grasses to decrease the remaining ATR significantly in the Alpha soil study II. Another reason for the different vegetation effects on the degradation of ATR and MET in the Alpha soil study II may be that the percentage of MET degraded in the

Table II. The influence of a mixture of three native prairie grasses on the degradation of metolachlor in Alpha and Bravo soils. Data are reported as percentage of remaining metolachlor at the end of each study

	Vegetation	*No Vegetation*	*Pr > F*	*Standard Error of Mean*
Alpha Soil Study I	37	77	0.0001	3.00
Alpha Soil Study II	51	74	0.0013	2.68
Alpha Soil Study III	63	93	0.0394	5.98
Bravo Soil Study	50	67	0.0008	0.03

rhizosphere of the plants is significantly greater than for ATR. Rice et al. indicated that MET was more readily degraded than ATR in water containing live aquatic plants (*27*).

The concentration of MET before vegetation was much greater than that of ATR in the Alpha soil study III. This is probably the reason why vegetation significantly decreased the concentration of MET, but not ATR in that study. The number of indigenous MET-degraders in Bravo soil is very low (*15*). That may explain the different effects of vegetation on the degradation of ATR and MET in the Bravo soil study.

NH_4NO_3 had no effect on the degradation of ATR and MET both in the Alpha soil study I (P = 0.9404 for ATR and P = 0.0734 for MET) and in the Bravo soil study (P = 0.8049 for ATR and P = 0.7461 for MET). In the N-amended soils, the remaining ATR and MET was 11% and 61%, respectively, while 12% of ATR and 52% of MET remained in the nonamended soils in the Alpha soil study I. In the Bravo soil study, the remaining ATR and MET was 13% and 58%, respectively, in the N-amended soils, while 13% of ATR and 59% of MET remained in the nonamended soils. This indicates that NH_4NO_3 did not influence the degradative ability of indigenous ATR or MET degraders in either soil. Our results are consistent with the findings of Entry, who noted that degradation of ATR by indigenous ATR-degraders was not affected by the addition of 400 kg N ha^{-1} in blackwater soils (*28*). However, others reported that exogenous N suppressed the ATR degradation by the indigenous ATR-degraders (*29-31*).

Conclusions

The degradation of ATR by the mixture of native prairie grasses was influenced by the concentration of ATR before vegetation, the presence of indigenous ATR-degraders, and the duration of the growth of the grasses in soil. The grasses significantly decreased the MET residues in both Alpha and Bravo soils. The enhanced degradation shown in our results suggests that phytoremediation with native prairie grasses could provide an inexpensive, effective, and aesthetically pleasing way to remediate soils contaminated with high concentrations of ATR and MET.

Acknowledgements

Funding for this research was provided by Novartis Crop Protection (Greensboro, NC). We thank Jennifer Anhalt, Karin Tollefson, Brett Nelson, John Ramsey, and Piset Khuon for their technical support. This is journal paper

No. J-14298 of the Iowa Agriculture and Home Economics Experiment Station, Ames, Iowa 50011.

References

1. Goolsby, D. A.; Thurman, E. M.; Kolpin, D. W. *Geographic and temporal distribution of herbicides in surface waters of the upper Midwestern United States, 1989-90;* 91-4034; U.S. Geological Survey Water-Resources Investigation Report: Denver, CO, 1990; pp 183-188.
2. Hall, J. K.; Mumma, R. O.; Watts, D. W. *Agric. Ecosyst. and Environ.* **1991,** *37,* 303-314.
3. Thurman, E. M.; Goolsby, D. A.; Meyer, M. T.; Mills, M. S.; Pomes, M. L.; Kolpin, D. W. *Environ. Sci. Technol.* **1992,** *26,* 2440-2447.
4. Muir, D. C.; Baker, B. E. *J. Agric. Food Chem.* **1976,** *24,* 122-125.
5. Rostad, C. E.; Pereira, W. E.; Leiker, T. J. *Biomedical and Environmental Mass Spectrometry* **1989,** *18,* 820-827.
6. Jayachandran, K.; Steinheimer, T. R.; Somasundaram, L.; Moorman, T. B.; Kanwar, R. S.; Coats, J. R. *J. Environ. Qual.* **1994,** *23,* 311-319.
7. Burken, J. G.; Schnoor, J. L. *J. Environ. Engineering* **1996,** *122,* 958-963.
8. Shimp, J. F.; Tracy, J. C.; Davis, L. C.; Lee, E.; Huang, W.; Erickson, L. E. *Crit. Rev. Environ. Sci. Technol.* **1993,** *23,* 41-77.
9. Schnoor, J. L.; Licht, L. A.; McCutcheon, S. C.; Lee-Wolfe, N.; Carreira, L. H. *Environ. Sci. Technol.* **1995,** *29,* 318A-323A.
10. Ryan, J. A.; Bell, R. M.; Davidson, J. M.; O'Connor, G. A. *Chemosphere* **1988,** *17,* 2299-2323.
11. Paterson, S.; Mackay, D.; Tam, D.; Shiu, W. Y. *Chemosphere* **1990,** *21,* 297-331.
12. Briggs, G. G.; Bromilow, R. H.; Evans, A. A. *Pestic. Sci.* **1982,** *13,* 495-504.
13. Walker, A.; Featherstone, R. M. *J. Exp. Botany* **1973,** *24,* 450-458.
14. Alvey, S.; Crowley, D. E. *Environ. Sci. Technol.* **1996,** *30,* 1596-1603.
15. Arthur, E. L.; Perkovich, B. S.; Anderson, T. A.; Coats, J. R. *Water, Air, and Soil Pollution* **2000,** *119,* 75-90.
16. Scheunert, I.; Qiao, Z.; Korte, F. *J. Environ. Sci. Health* **1986,** *B21,* 457-485.
17. Roeth, F. W.; Lavy, T. L. *Weed Sci.* **1971,** *19,* 93-97.
18. Anderson, T. A.; Guthrie, E. A.; Walton, B. T. *Environ. Sci. Technol.* **1993,** *27,* 2630-2636.
19. Walton, B. T.; Guthrie, E. A.; Hoylman, A. M. In *Bioremediation Through Rhizosphere Technology*; Anderson, T. A.; Coats, J. R., eds.; ACS Sym. Ser. 563; American Chemical Society: Washington, DC, 1994; pp 11-26.

20. Zablotowicz, R. M.; Locke, M. A.; Hoagland, R. E. In *Phytoremediation of Soil and Water Contaminants;* Kruger, E. L.; Anderson, T. A.; Coats, J. R., eds.; ACS Sym. Ser. 664; American Chemical Society: Washington, DC, 1997; pp 38-53.
21. Ferro, A. M.; Sims, R. C.; Bugbee, B. *J. Environ. Qual.* **1994,** *23,* 272-279.
22. Perkovich, B. S.; Anderson, T. A.; Kruger, E. L.; Coats, J. R. *Pestic. Sci.* **1996,** *46,* 391-396.
23. Crowley, D. E.; Alvey, S.; Gilbert, E. S. In *Phytoremediation of Soil and Water Contaminants;* Kruger, E. L.; Anderson, T. A.; Coats, J. R., eds.; ACS Sym. Ser. 664; American Chemical Society: Washington, DC, 1997; pp 20-36.
24. Hoagland, R. E.; Zablotowicz, R. M.; Locke, M. A. In *Phytoremediation of Soil and Water Contaminants;* Kruger, E. L.; Anderson, T. A.; Coats, J. R., eds.; ACS Sym. Ser. 664; American Chemical Society: Washington, DC, 1997; pp 92-105.
25. Wetzel, S. C.; Banks, M. K.; Schwab, A. P. In *Phytoremediation of Soil and Water Contaminants;* Kruger, E. L.; Anderson, T. A.; Coats, J. R., eds.; ACS Sym. Ser. 664; American Chemical Society: Washington, DC, 1997; pp 254-262.
26. Struthers, J. K.; Jayachandran, K.; Moorman, T. B. *Appl. Environ. Microbiol.* **1998,** *64,* 3368-3375.
27. Rice, P. J.; Anderson, T. A.; Coats, J. R. In *Phytoremediation of Soil and Water Contaminants;* Kruger, E. L.; Anderson, T. A.; Coats, J. R., eds.; ACS Sym. Ser. 664; American Chemical Society: Washington, DC, 1997; pp 133-151.
28. Entry, J. A. *Biol. Fertil. Soils* **1999,** *29,* 348-353.
29. Ames, R. A.; Hoyle, B. L. *J. Environ. Qual.* **1999,** *28,* 1674-1681.
30. Alvey, S.; Crowley, D. E. *J. Environ. Qual.* **1995,** *24,* 1156-1162.
31. Gan, J.; Becker, R. L.; Koskinen, W. C.; Buhler, D. D. *J. Environ. Qual.* **1996,** *25,* 1064-1072.

Chapter 10

Persistence, Mobility, and Bioavailability of Pendimethalin and Trifluralin in Soil

J. B. Belden[1], T. A. Phillips[1], K. L. Henderson[1], B. W. Clark[1], M. J. Lydy[2], and J. R. Coats[1]

[1]Department of Entomology, Iowa State University, 116 Insectary, Ames, IA 50011–0001
[2]Department of Zoology, University of Southern Illinois, Carbondale, IL 62901–5601

Pendimethalin and trifluralin are current-use pesticides that have been previously reported as persistent, bioaccumulative, and toxic. In the studies presented here, dissipation of aged and fresh residues of pendimethalin and trifluralin were evaluated in soil, as well as the bioavailability of residues to earthworms and the movement of pendimethalin in a soil column. In a separate study, pond water receiving runoff from a golf course was measured for the presence of pendimethalin. Dissipation measurements of pendimethalin and trifluralin in soil indicated very slow dissipation with 40–60% of the compounds extractable at 1026 days after the first measurement. In a second study, dissipation of pendimethalin was more rapid, however more than 30% was present after 310 days of soil treatment. Biovailability, as measured by earthworm biological accumulation factors, was reduced over time. Mobility of pendimethalin was very limited. Almost no downward movement was measured in the column study, and no detectable levels were found in runoff from turf grass.

Introduction

Dinitroanaline herbicides are used for pre-emergent control of grass weeds in many crops including soybeans, cotton, and turfgrass. In plants, these herbicides bind to tubulin, thus preventing microtubule formation, disrupting mitosis, and causing ultrastructural effects (1). Throughout the last thirty years, large amounts of both of these herbicides have been applied to soil. In 1996 alone, five million kilograms of trifluralin and six million kilograms of pendimethalin were applied to soybean fields in the United States (2). Herbicide use can cause nonpoint source contamination of surface and groundwater. Additionally, due to the large amount of herbicide usage, incidental spillage during mixing and loading can occur, resulting in heavily contaminated soil (3).

Previous studies have indicated that soil found at agrochemical dealerships in Iowa had concentrations of pendimethalin above 100 mg/kg and concentrations of trifluralin above 10 mg/kg (4), levels well above application rates. Since the degradation rate of some pesticides is concentration-specific, with higher concentrations degrading more slowly (5, 6), the environmental impact of these sites may not be well predicted by current fate studies, which are mostly conducted using field conditions.

Pendimethalin and trifluralin have low water solubilty and sorb to soil at a high rate. However, some studies have indicated that pendimethalin and trifluralin may contaminate groundwater and surface water. For instance, pendimethalin has been found in groundwater near agrochemical dealerships (3) and both trifluralin and pendimethalin have been detected in surface water in studies conducted by the United States Geological Survey (7).

The presence of herbicides in surface and groundwater is widely studied because of the potential impact on human and environmental health. Although less studied, soil contamination can also impact the environment in a variety of ways. First, the contaminant can move into aqueous compartments by leaching into groundwater or entering surface water through runoff. Secondly, the chemical can cause local effects due to direct toxicity to plants or animals. Finally, the contaminant can accumulate in local fauna or flora and cause toxicity to animals at higher trophic levels.

Both of the most intensively used dinitroanaline herbicides, pendimethalin and trifluralin, have been placed on the United States Environmental Protection Agency's list of Persistent Bioaccumulative and Toxic chemicals (PBT; 8). Most of this list is composed of industrial and past-use pesticides such as chlorinated insecticides, mercury, polychlorinated biphenyls, polyaromatic hydrocarbons, and DDT. The dinitroanalines are the only compounds on the PBT list that are currently produced in the United States for release into the environment at high levels, although the environmental fate and effects of dinitroanilines have been less studied as compared to other PBTs.

In the past few years we have conducted a variety of studies to determine the fate of herbicides in soil. Each study investigated several herbicides and was designed to evaluate either best management practices for reduction of pesticide runoff or phytoremediation of herbicide residues in soil. The portions of each study regarding pendimethalin or trifluralin movement and dissipation are summarized within this manuscript to provide an overview of the environmental fate data we have collected regarding dinitroanaline herbicides. Complete studies reporting data for all pesticides involved and all remediation strategies will be published elsewhere.

Methods

Microplot Study

Soil from an agrochemical dealership site (loamy sand, 1.6% organic matter) was treated with 100 ppm atrazine, 25 ppm metolachlor, and 25 ppm trifluralin in addition to the 110 ppm of pendimethalin and 10 ppm metolachlor already present. The soil was distributed between four containers (24 x 30 cm base with 18-cm depth). Microplots were placed outside during the summer months (near a cornfield on the Iowa State University campus) and within a greenhouse during winter months (20°C, 16:8 light to dark). Soils were aged for thirty days before the first sampling. Plots were sampled by taking three soil cores at various points in time up to 1026 days. Concentration of the pesticides was measured at each time point. After 1026 days, soil from each plot was allowed to dry and then mixed thoroughly. Chemical analysis and soil earthworm bioaccumulation studies were performed to evaluate the bioavailability of the remaining dinitroanaline residues.

Extraction and Analysis of Soils

Soils were extracted as previously reported (9). Briefly, 20 g of soil were shaken three times with 60 ml ethyl acetate. The resulting extract was concentrated either under nitrogen flow or by a rotary evaporator. Analysis of the extracts was performed by gas chromatography and thermionic specific detection (GC-TSD) as previously described (9).

Earthworm Bioassay

Eight-day earthworm bioaccumulation assays were conducted as previously reported (10). Four adult worms (*Eisenia fetida,* average mass 1.5 g) were placed in 200-ml jars with 150 g of the test soil. Soil moisture was adjusted to 19% (approximately 1/3 bar) before sealing the jar with perforated Parafilm®. After eight days of incubation at 25°C and constant low level light, earthworms were removed from the sample soil and placed in untreated soil for one day. Worms were then removed and extracted three times with 10 ml of ethyl acetate in a sample homogenizer. Measurement of trifluralin or pendimethalin was then performed using GC-TSD. Biological accumulation factors (BAFs) were calculated as concentration in the earthworm divided by the concentration recoverable by ethyl acetate extraction, and these BAFs were used as a measure of bioavailability. Control studies were performed identically using soil that had been fortified with herbicide 24 hours before the start of the test. Control studies provide a baseline for the expected BAFs for pendimethalin and trifluralin for fresh residues.

Soil Column Study

Eight soil columns were constructed in PVC pipe (10 cm in diameter, 23 cm long) enclosed at the bottom with aluminum screen and glass wool. The soil used was collected from a corn field near Ames, IA that had not received herbicide treatment (sandy loam, 2.7% O.M.; Field 55, ISU Ag Engineering/Agronomy Farm). The columns were packed with 7 cm of unfortified soil at the base of the column, and topped with 14 cm of soil fortified with atrazine, alachlor, metolachlor, and pendimethalin each at 25 mg/kg (36.9 mg per column). Columns were packed to a bulk density of 1.25 g/ml. Distilled water was added slowly (1 cm/hour) to each column to reach a moisture content of 1/3 bar (19.1%). Columns were maintained at greenhouse conditions (25°C, 16:8 minimum light to dark). Each column was watered with 1 cm (81 ml) of distilled water every 96 hours. This amount kept the columns moist, yet was insufficient to cause loss of water from the bottom of the column.

Leaching of Soil Columns and Removal of Soil

After 240 and 330 days, 7.5 cm (608 ml) very soft water (12.0 mg/L as NaHCO3) was added to each column over an 8-hour period. Leachate was collected at the bottom of the column. Leachate was stored at 4°C no longer than 48 hours before analysis by solid phase extraction. Extracts were analyzed by GC-TSD. The columns were then left at greenhouse conditions without water for ten days after the leaching process (250 and 340 days). Soil from each column was collected from three regions: the top 6 cm, the following 6 cm (middle), and the bottom 5 cm. The soil was sieved (2.8 mm) and then stored in glass containers at 4°C. Chemical analysis and earthworm bioassays were conducted as previously discussed.

Golf Course Study

Water samples were collected from Braeburn Golf Course in Wichita, KS. The course is an 18-hole, 160-acre, public golf course that contains four ponds. Pendimethalin was applied as a pre-emergent herbicide on the fairways of the course throughout the sampling period. The formulation applied was a fertilizer and pendimethalin mixture (Scott's Brand) applied at 134 kg/ha (active ingredient 1.2%, 1.6 kg/ha). Samples were collected from two ponds. Pond A receives runoff from the southern and western portions of the course, as well as flow from a pond not studied, and flow from drainage tiles. Pond A is small (800 m^2) and shallow (<0.3 m deep). Pond B receives flow from fairways and from an unstudied pond. It is a larger (3000 m^2) and deeper (average 2.5 m) pond. Water draining from Pond B leaves the golf course and enters the Arkansas River Watershed. A more thorough description of the site and sampling techniques is available elsewere (11).

The study lasted for three years, from July 1997 to October 2000. Samples were taken from the two ponds biweekly. In addition, water was collected during rain events five times in the first two years and ten times in the third year of the project. Rainfall had to exceed 1.5 cm to qualify as a rain event. During each rain event, samples were collected 2, 4, and 24 hours following the onset of rain. Samples were collected approximately 0.5 m from the shoreline as grab samples in 1 L glass bottles. Samples were transported to the laboratory and analyzed within 48 hours.

Pesticides were extracted by solid-phase extraction and analyzed by gas chromatography coupled with nitrogen-phosphorus detection as previously described (12). Samples were fortified with tributylphosphate in acetone as a surrogate. Recovery for pendimethalin using this method was 80.5 % with a standard deviation of 6.8 % (n = 36). Quantitation limits were 0.4 µg/L.

172

Results

Microplot Study

Pendimethalin and trifluralin had very low dissipation rates. As shown in Figure 1, both compounds had very little dissipation after the first thirty days of aging. Only small losses occurred throughout the first year, and greater than 50% of each compound was present after 1000 days. At the end of the study, a higher percentage of pendimethalin had dissipated as compared to trifluralin, even though it was present in the soil at higher concentrations and was aged for a longer period prior to the beginning of measurement collection.

Pendimethalin and trifluralin accumulated in earthworms. After eight days of exposure to soil from the microplots, BAF values were calculated as 2.9 (SE 0.2) for trifluralin and 0.78 (SE 0.08) for pendimethalin. BAF values calculated for exposure in control soil (as reported for the column study) aged for only one day were 5.7 (SE 0.5) for trifluralin and 1.9 (SE 0.2) for pendimethalin. For both pendimethalin and trifluralin, the BAFs were significantly lower in microplot soil as compared to freshly treated soil (p < 0.01).

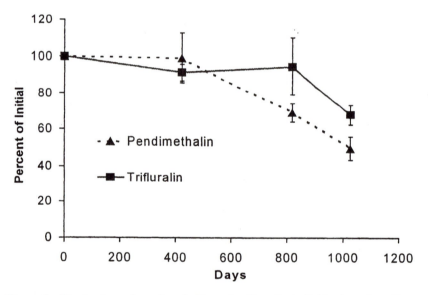

Figure 1. Dissipation of pendimethalin and trifluralin in field microplots. Error bars illustrate one standard error.

Column Study

Dissipation of pendimethalin in the study involving soil columns also was slow. After 250 days in the soil column, 41% (SE 2) was present, and after 340 days, 31% (SE 2) was present. As shown in Figure 2, dissipation was significantly faster in the top of the column than in the middle section (p < 0.05) and was greater as time increased (p < 0.05). In addition, pendimethalin moved very little in the soil column. Pendimethalin was not detectable in the leachate (less than 0.05% of amount applied), and less than 1.5% of the applied amount moved into the bottom unfortified soil section.

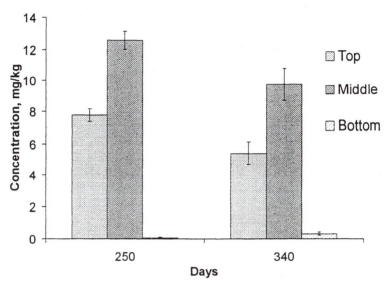

Figure 2. Concentration of pendimethalin in soil column sections following aging and leaching. Twenty-five mg/kg of pendimethalin was applied to the top two sections of the column. Length of time in the soil and section position both significantly affected the amount of pendimethalin remaining in the soil.

Pendimethalin also was detected in earthworms exposed to soil from the top and middle sections of the columns. As shown in Figure 3, bioavailability dropped in all sections and times as compared with the BAF values calculated for exposure in control soil (1.9, SE 0.2). BAFs were significantly lower for the top section as compared with the middle section (p <0.05) and for columns aged 340 days as compared with 250 days.

Golf Course Study

Pendimethalin was not detected in any of the samples collected in this study. Other compounds, such as simazine, which were applied to the fairways, were detected in this study as previously reported (11). Quality control for the analytical procedures demonstrated the effectiveness of the measurement method. Matrix spike and surrogate recoveries were above 70% for all samples.

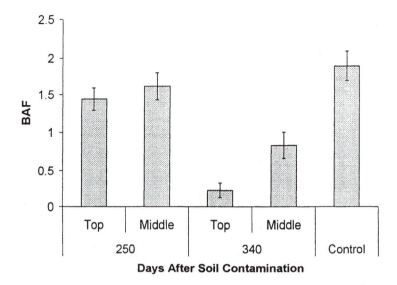

Figure 3. Biological accumulation factors (BAFs) for pendimethalin accumulation in earthworms. Greater length of time in soil and location in the top of the soil column significantly decreased BAF values.

Discussion

These results demonstrate the persistence of pendimethalin and trifluralin. At the end of the microplot study, less than 50% dissipation had occurred for both compounds, indicating that the half-life of the residues was greater than 1000 days. In the column study, the dissipation rate of pendimethalin was faster, yet after 340 days, 31% still remained. Although the number of time points was limited, the half-life of the residues could be estimated as greater than 170 days, as an average of the first two half-lives. The difference in pendimethalin dissipation rates between the studies may be the result of several factors. First, the agronomic soil used in the column study had higher organic matter, higher clay content, and may have been more biologically active than the sandy, previously contaminated soil used in the microplot study. But, increases in organic matter content usually result in the reverse trend with greater adsorption to the soil. Second, in the column study, soil was kept at 25°C and at moisture conditions close to the gravimetric water potential throughout the study, while in the microplot study, soil was left outdoors, subject to temperature and moisture extremes. At times, these conditions were likely to be cooler and dryer than ideal for degradation. Pendimethalin has been shown to dissipate more slowly with decreased soil moisture (14). Finally, pendimethalin within the microplots had been aged for an unknown amount of time before the start of the test, and therefore may not have been available for volatilization or biodegradation.

Previous studies have often reported shorter dissipation half-lives for pendimethalin and trifluralin. For instance, pendimethalin has been reported to have dissipation half-lives of 37 days in onion fields (13), 47 days in agronomic soil (14), and 12 days in thatch on a golf course (15). Other laboratory studies have reported pendimethalin half-lives as long as 98 days at 30°C and 407 days at 10°C (16). Trifluralin residues have been shown to persist in fields, with half-lives of 35.8 and 25.7 days (17) or longer (18). However, additional studies have indicated that after a period of initial dissipation, both compounds may have much slower dissipation rates (17, 18). Concern has been raised about the validity of using dissipation rates as a measurement of persistence. Degradation half lives for both compounds may be much longer (8). But, dissipation rates were measured in these studies and, in our opinion, are relevant to soil concentrations in the field.

The dissipation rates found in both the microplot and column studies could be slow for a variety of reasons. Both trifluralin and pendimethalin may have dissipated due to volatility and photodegradation (18, 19, 20). Rapid incorporation of the pesticides into the soil during these studies may have decreased the magnitude of these important dissipation mechanisms. The concentrations used in these studies were also higher than those used in previous studies, in order to represent point-source spillage instead of field application. Increased concentration has been shown to decrease degradation rates for some compounds (5, 6).

Pendimethalin was shown to have minimal mobility in surface runoff and leachate in this study, as demonstrated by the lack of movement in the soil column study, and the less-than-detectable levels found in the ponds in the golf course study. These results are not unexpected. In general, the dinitroanalines have low water solubility and high soil adsorption (18). Field studies of pendimethalin have indicated very little leaching through soil columns (21). This lack of movement, coupled with previous reports of higher dissipation rates of pendimethalin on golf course thatch (15), may explain why detectable water contamination was not present on the golf course. However, if contamination of the water did occur, pendimethalin is likely to partition quickly into the sediment, which was not tested.

Overall, BAF values for pendimethalin and trifluralin in these studies decreased over time as compared to BAF values for fresh residues. This result was also expected. Many persistent contaminants have been shown to have decreased availability over time, and it has been suggested that the change in bioavailability should be considered when estimating risk (22). In the column study, the BAF values for the top section were significantly lower than for the middle section (p <0.05). Due to the overall method design of the experiment, potting soil rich in organic matter was added to the top of the column. This increase in organic matter may have resulted in the decrease in bioavailability.

In summary, our results indicate that pendimethalin and trifluralin may be very persistent herbicides when incorporated into soil. However, when considering the potential environmental hazards of these residues, one should consider the low mobility of the compounds within the soil and the decreased bioavailability that occurs as the residues age.

Acknowledgements

Partial financial support for this project was provided by the Center for Health Effects of Environmental Contaminants (CHEEC) at the University of Iowa. This chapter is publication No. J-19837 of the Iowa Agriculture and Home Economics Experiment Station, Project 3187. Financial assistance has also been provided by an EPA Section 319 Non-Point Source Pollution Control Grant C9007 405-97 through an agreement with the Kansas Department of Health and Environment. We would also like to mention John Wright golf course general manager, and superintendent.

References

1   Devine, M., S.O. Duke, and C. Fedtke. 1993. Microtubule disruptors. In *Physiology of Herbicide Action*. Prentice Hall. Englewood Cliffs, NJ: pp 190-225.

2   National Agricultural Statistics Service. 1997. Agricultural Chemical Usage, 1996. U.S. Department of Agriculture, Washington, D.C.

3   Gannon, E. 1992. Environmental Clean-up of Fertilizer and Agrichemical Dealer Sites - 28 Iowa Case Studies. Iowa Natural Heritage Foundation, Des Moines, IA.

4   Arthur E.L., and J. R. Coats. 1998. Phytoremediation. In *Pesticide Remediation in Soils and Water*. P.K. Kearney, T. Roberts Eds. John Wiley & Sons. Washington D.C.: 251-281.

5   Gan, J., R.L. Becker, W.C. Koskinen, and D.D. Buhler. 1996. Degradation of atrazine in two soils as a function of concentration. J. Environ. Qual. 25:1064-1072.

6   Gan, J., W.C. Koskinen, R.L. Becker, and D.D. Buhler. 1995. Effect of concentration on persistence of alchlor in soil. J. Environ. Qual. 24:1162-1169.

7   Gilliom, R.J., J.E. Barbash, D.W. Kolpin, and S.J. Larson. 1999. Testing water quality for pesticide pollution. Environ. Sci. Technol. 33: 164A-169A.

8   United States Environmental Protection Agency. 1999. Persistent bioaccumulative toxic (PBT) chemicals: final rule. 40 CFR Part 372.

9   Anderson, T.A., E.L. Kruger, and J.R. Coats. 1994. Enhanced degradation of a mixture of three herbicides in the rhizosphere of a herbicide-tolerant plant. Chemosphere. 28:1551-1557.

10 Kelsey, J.W., B.D. Kottler, and M. Alexander. 1997. Selective chemical extractants to predict bioavailability of soil-aged organic chemicals. Environ. Sci. Technol. 31: 214-217.

11 Davis, N., and M.J. Lydy. (2002). Evaluating best management practices at an urban golf course. Environmental Toxicology and Chemistry. 21: 1076-1084.

12 Belden, J.B., and M.J Lydy. 2000. Analysis of multiple pesticides in urban storm water using solid-phase extraction. Arch. Environ. Contam. Toxicol. 38: 7-10.

13 Tsiropoulos, N.G., and G. E. Miliadis. 1998. Field persistence studies on pendimethalin residues in onions and soil after herbicide postemergence application in onion cultivation. J. Agric. Food. Chem. 46:291-295.

14 Zimdahl, R.L., P. Catizone, and A.C Butcher. 1984. Degradation of pendimethalin in soil. Weed Sci. 32:408-412.

15 Horst, G.L., P.J. Shea, N. Christians, D.R. Miller, C. Stuefer-Powell, and S.K. Starrett. 1996. Pesticide dissipation under golf course fairway conditions. Crop Sci. 36:362-370.

16 Walker, A., and W. Bond. 1977. Persistence of the herbicide AC 92, 55, N-(1-ethylpropyl)-2,6-dinitro-3,4-xylidine in soils. Pestic. Sci. 8:359-365.

17 Duseja, D.R., and E.E. Holmes. 1978. Field persistence and movement of trifluralin in two soil types. Soil Sci. 125:41-48.

18 Helling, C.S. 1976. Dinitroanaline herbicides in soil. J. Environ. Qual. 5:1-15.

19 Dureja, P., and S. Walia. 1989. Photodecomposition of pendimethalin. Pestic. Sci. 25:105-114.

20 Schroll, R., U. Dorfler, and I. Scheunert. 1999. Volatilization and mineralization of ^{14}C-labelled pesticides on lysimeter surfaces. Chemosphere. 39:595-602.

21 Zheng, S.Q., J.F. Cooper, and P. Fontanel. 1993. Movement of pendimethalin in soil of the south of France. Bull. Environ. Contam. Toxicol. 50:492-498.

22 Kelsey, J.W., and M. Alexander. 1997. Declining bioavailability and inappropriate estimation of risk of persistent compounds. Environ. Toxicol. Chem. 16:582-585.

Chapter 11

Evaluation and Effective Extraction Method of Paraquat Residue of Soil in Korea

Kyu Seung Lee[1] and Jin Wook Kwon[2,3]

[1]Department of Agricultural Chemistry, Chungnam National University, Yusong, Taejon, 305–764, Korea
[2]Toxicology Research Center, Korea Research Institute of Chemical Technology, Yusong, Taejon, 305–6000, Korea
[3]Current address: National Veterinary Research and Quarantine Service, 620–2 Amnam-dong, seo-gu, Busan 602-833, Korea

Abstract

Paraquat, a contact herbicide that has been widely used in Korea, was reviewed for its use pattern and social issues brought forward environmental and consumer groups. Orchard soils in Korea were monitored for two years, and paraquat was detected in more than 95% of the samples, and the average residue level was 8.1 mg/kg, in 1996 and 6.87 mg/kg in 1997, respectively. At the same time, SAC-WB, regarding as safety margin of paraquat adsorption in soil, were carried out wheat bioassay at the equilibrium concentration and was observed to the same soil sample. Average SAC-WB value were 223.40 mg/kg in 1996 and 296.12 mg/kg in 1997, respectively. An extraction method for paraquat residue in soil was looked over to improve reproducibility and efficiency of recovery. Pre-washing with concentrated HCl for 1 hr. is suggested for that purpose. Finally, capillary zone electrophoresis was used for the detection of low amounts of paraquat residue as level of $10^{-13} \sim 10^{-14}$ g/l in paddy water.

Paraquat (1,1-dimethyl-4,4'-bipyridylium dichloride), a para-substituted quaternary bipyridyl cation, was introduced commercially in 1962 as a contact herbicide and has been used in Korea since 1970. It exists as very strong soluble chloride salts in dry conditions, and very stable bipyridyl cation in aqueous environments [1]. These characteristics allow for foliar application. Paraquat was strongly bound to soil clay minerals such as montmorillonite and other negatively charged soil components, including organic matter [2].

Paraquat is inactivated when adsorbed, therefore no translocation through the roots and no harmful effects to following crops were found after paraquat application to the soil. Because degradation by microorganisms and chemical processes were not especially effective [3, 4], we assume that the paraquat residues can continuously increase and accumulate in the soil [3].
Nevertheless, the residue levels of paraquat in soil after continuous use in the same soil were not much higher than expected [5,6], while paraquat residue levels decreased in soil after application stopped [7]. These results suggest degradation by microorganisms or other routes such as photo-decomposition. In fact, monopyridone, an oxidized metabolite of paraquat was found during degradation of plant residues [4] and was also identified in fungal-degradation studies[8]. The strong binding property of paraquat was mainly dependent on the kinds of clay mineral, clay content, and organic matter. Therefore, residual forms of paraquat where divided into two categories, loosely bound and tightly bound paraquat [9]. The former is desorbed by saturated ammonium chloride solution, and the latter by refluxing with conc. H_2SO_4. Tightly bound paraquat residues can not be extracted easily from soil, and they did not translocate or diffuse into the environment. Plants can not take up these residues in soil, and soil microbes cannot be affected even in the kaolinite-dominated soils which favor strong cation adsorption, although kaolinite shows less cation adsorption than other clay minerals. Various methods to determine of paraquat residues have been introduced like spectro-photometry [10 – 15], gas chromatography [16], HPLC [17, 18], ion-pair chromatography [19], enzyme-linked immunosorbent assay(ELISA) [20] and capillary electrophoresis [21-23]. Current extraction methods have low recovery, are time consuming, and give relatively high errors in reproducibility. Therefore, an effective extraction method of paraquat residue in soil needs to be developed for evaluation and risk assessment of paraquat- treated soil.

Status of paraquat use in Korea

Paraquat has been used for weed control in orchards and non-crop land since it was introduced in Korea in 1970. Since then, the consumption rate increased until 1995[24] (Figure 1). The ratio of paraquat amounts per the applicable area which include orchard and mulberry cultivation area were calculated. (Figure 2) Three applications of the recommended dose per applicable area per year is regarded as a normal spray pattern and a standard application amount. In Korea, 56% of the total paraquat use was for upland crops, 4% in direct rice, 6% in pre-planting rice and 5% in barely[25], and 44% for non-crop areas, including 28% in rice and upland bund, and 16% in other non-agricultural use. However, the total consumption amounts of paraquat were 3~4 times more than that of estimated from standard application to the applicable areas. Therefore, the environmental and consumer groups indicate that paraquat residues in the soil may be hazardous by steady accumulation and excess of the maximum holding capacity of soil in near future. Moreover, paraquat is notorious for being drunk in suicide attempts, leading consumer groups to insist on the banning of paraquat application because of toxicity to humans. A report from June, 1992 to October, 1996 in Korea [26] found that the paraquat intoxication ratio was 36.4%, slightly higher than that of total organophosphorus insecticides, 35.6%. Hence, the National Regulatory Committee of pesticides decided to manage the special review for the environmental safety of paraquat during 1996 ~ 1997 including: 1) Current levels of paraquat residues in the orchard soil and products, and 2) biological effects of paraquat use on crops that follow.

Evaluation of paraquat residue in soil

Sixty orchard soils were observed in 1996 and 1997, respectively. Twenty soils were from apple, 15 from pear, 12 from grape and 13 from peach orchards in 1996. Also, 15 soils from each orchard were sampled in 1997 (Table I).

The loosely bound residues of paraquat in soil were also determined for all samples, and residue amounts were below detectable limit in all of tested soils.

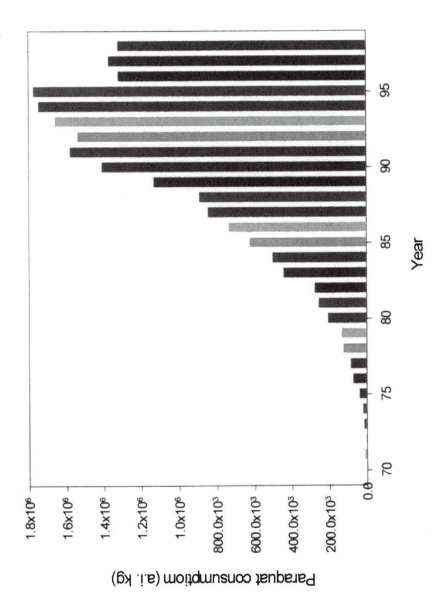

Fig. 1. Consumption of paraquat in Korea.

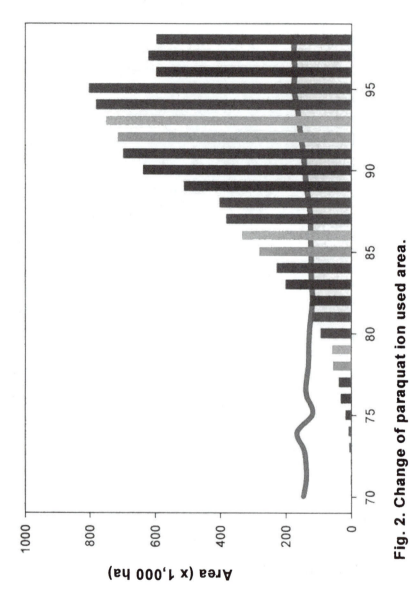

Fig. 2. Change of paraquat ion used area.
The area below red line is permitted to paraquat application area which equals to 3 times application per year.

Table I. Evaluation of Paraquat residue in orchard soil during 1996 ~ 1997 [27, 28]

Year	Orchard	Occurrence of frequency	Paraquat Residue(mg/kg)		
			Minimum	Maximum	Average±SD
1996	Apple	18/20	<0.05	30.02	7.91±7.03
	Pear	15/15	1.50	23.81	7.57±6.96
	Grape	12/12	0.76	35.02	9.99±10.63
	Peach	13/13	3.50	20.85	7.25±4.79
	Average	58/60 (96.7%)			8.10±7.44
1997	Apple	15/15	3.64	20.18	10.26±4.62
	Pear	14/15	<0.5	11.17	5.58±3.54
	Grape	15/15	2.57	10.43	5.38±2.34
	Peach	15/15	2.87	9.73	6.27±2.35
	Average	59/60 (98.3%)			6.87±3.86

Parallel with the monitoring survey, SAC-WB (Strong Adsorption Capacity-Wheat Bioassay) were carried out for the soil samples. Strong Adsorption Capacity (SAC) is defined as the paraquat concentration that is adsorbed strongly by soil at the equilibrium concentration. Through the wheat bioassay (WB) [17,29] we can find the 50% inhibition concentration with 14 day-old wheat root. This bioassay method is useful to find the adsorption capacity for safe levels of paraquat. Actually, although paraquat residue reached the SAC-WB concentration or more, we did not observe any phytotoxic effects.

The relationship between paraquat residues or SAC values and some soil properties was observed. Residue levels of paraquat were not significantly correlated to clay content, CEC or organic matter of soils. The only significant observation was the SAC-value vs. clay content (Figure 3). Residue ratio per SAC-WB + soil residue was observed in accordance with clay contents. (Figure 4) Less than 16% of clay content is regarded as loamy sand, 17 ~ 21% as sandy loam, 22 ~ 29% as loam and more than 30 % as clay loam or silt clay loam. This classification is not exactly the same as the soil taxonomy system of the USDA [30]. From Figure 4, we found that lower clay content in soil showed less SAC-WB and more residual rate. In loamy sand soil, SAC-WB ranged from 88.4 to 126 mg/kg and 7.0 to 9.9% of SAC-WB + residue was observed in residue. However, it ranged from 321.6 to 377 mg/kg and 1.7 to 2.9% in silt clay loam and clay loam soil respectively. The more clay content of soil, the higher the observed SAC value.

Half-life of paraquat in Korean soil was also investigated. The results of half-life experiments with various soil conditions are shown in Table III. The half-life of paraquat in soil under field conditions is quite long, such as 118 to 347 days in a single application, and 158 to 347 days in a duplicate application within one week interval. The more clay content in soil, the longer the half-life, but organic matter content did not influence half-life. The results from 1998 showed much longer half-lives than in 1994, however application time differed between the two experiments. Paraquat is more likely to be lost in run-off of soil particles by heavy rainfall, and degradation by active microorganisms is likely to occur in early May applications. On the contrary, application in late August should not be affected by these factors as much as in early May under normal Korean weather conditions.

Table II. SAC-WB for orchards soils

Year	Orchard	No. of sample	SAC-WB (mg/kg)		
			Minimum	Maximum	Average ± SD
1996	Apple	20	113	425	210 ± 92
	Pear	15	105	422	258 ± 94
	Grape	12	139	465	242 ± 100
	Peach	13	78	316	182 ± 70
	Average	60	78	465	223 ± 89
1997	Apple	15	75	478	262 ± 130
	Pear	15	58	392	267 ± 102
	Grape	15	57	650	298 ± 187
	Peach	15	37	699	276 ± 193
	Average	60	37	699	276 ± 153

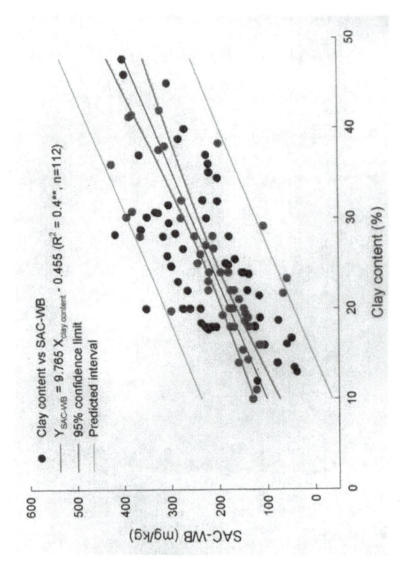

Fig 3. The relationship between SAC-WB and soil clay contents.

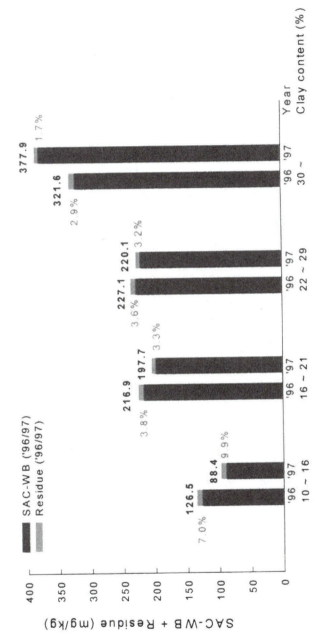

Fig. 4. The relationship between SAC-WB and residue ratio in accordance to clay contents. Numerics are SAC-WB + Residue, % numerics are the ratio of residue per SAC-WB + Residue.

Table III. Half-life of paraquat in various soils.

Soil texture	Clay content (%)	O.M (%)	Application [1]	DT50(day)	Remark
Sandy loam	16.5	0.6	1	118	Lee et al. (1994) [31]
			2 [2]	169	
Loam	25.4	1.7	1	131	
			2	158	
Clay loam	24.2	0.9	1	347	Kim et al. (1998) [32]
			2	347	
Loam	11.5	1.9	1	203	
			2	330	

1) 24.5% paraquat soln. applied at 6L/Ha basis (0.147 kg A.I./10a)
2) Duplication within 1 week interval

Determination of paraquat in soil

Some differences in the analytical process between Zeneca and Korea.

Determination methods of paraquat residues in soil, water and other biological samples have been provided by Zeneca [33 ~ 35]. Modifications of Zeneca methods may provide more accurate measurement and higher recovery of paraquat. Basic analytical procedures are : soil samples are extracted by refluxing with sulfuric acid; digested samples are filtered; column clean-up with cation exchange resin is used for fractionation of paraquat; a portion of column eluent is treated with a strong reducing agent and then measured by spectrophotomety. However, some steps of the analytical procedure are slightly different including extraction and column clean-up.

Table IV shows the differences between Zeneca's recommended method and the Korean standard method.

Although the differences are minor, several of the steps of the paraquat residue analysis procedure are not the same. In the filtration step, Zeneca recommends the digest of sulfuric acid be diluted to 1,000 ml with de-ionized H_2O. On the contrary, Korean standard method emphasizes the removal of any inhibiting cations contained in the digest by addition of EDTA-2Na at pH 7.0 and increasing the pH of solution to 8-9 for aiding the filtration velocity. Also, Zeneca recommended saturated NH_4Cl for desorption of the paraquat ion from the cation resin, while the Korean standard method recommended 50% NH_4NO_3 for the same purpose. Some metal ions precipitate as inactive metal chlorides by reacting with the ammonium chloride eluent. NH_4NO_3 does not make any precipitated form by ionic competition. Therefore, EDTA-2Na must be used to remove the metal ions in the case of ammonium nitrate eluents.

Effective extraction of paraquat residue in soil

Current methods used to measure paraquat residues are time consuming and result in poor recovery. During the filtration procedure, the extracted paraquat ion might be readsorbed onto clay minerals. When

Table IV. Different points between Zeneca and Korean standard methods.

	Zeneca	Korean standard	Remark
Sample size	25 g	50 g	
Concentration of sulfuric acid	6 M	9 M	
Filtration of digest	Dilute to 1,000 ml and filter	Adjust pH to 7.0 add 5% EDTA-2Na readjusting pH to 8-9 and filter	Removes other interferring cations, such as Zn, Cu, Fe *etc.*
Final eluent of cation exchange resin column chromatograph	Saturated NH_4Cl	50% NH_4NO_3	NH_4Cl may make chloride precipitate with metal ions

NaOH is used for neutralization purpose as the Korean standard method, the digested solution changes to sticky paste. Hence, it is possible for some paraquat residues to remain on the filtrate as a re-adsorbed form. At the same time, the sticky paste form usually causes an overdose of NaOH solution because it is difficult to find the neutralization point. This phenomena has caused a long filtration time, because of the large volume, and increasing the re-adsorption possibility during extraction. This phenomenon is caused by the dispersion of Na^+ and other cations from the destroyed clay mineral through strong acid hydrolysis, although some soluble portions of organic matter could be contributing to this condition. Therefore, to solve this problem, it is necessary to improve the recovery rate and reduce the analytical time. For this purpose, pre-washing with conc. HCl showed remarkable improvement of existing methods. Conc. HCl can remove a soluble organic matter and some exchangeable cations which may disperse into acidic solution of conc. H_2SO_4 digestion.

The loosely bound residue, an ionic form of paraquat, may be removed by pre-washing with conc. HCl. When 1 and 10 mg/kg level of paraquat were treated with soil and the amount of loosely bound residue was checked, it was below the limit of quantitation (LOQ) of recommended analytical method.[36] It means that loosely bound residue is not much affect to the residue level of paraquat in soil. Effects of pre-washing with conc. HCl were compared on the view point of shaking time course (Table V)

As shown in Table VI, 1 hour of shaking with 70 ml of conc. HCl increased recovery for soil A 1.15 times and 1.09 times for soil B compared to controls. Also, the recoveries of paraquat from the two soils could be affected by clay contents. At the same time it was found that filtration time was also reduced 4.64 times for soil A and 4.91 times for soil B. Improvement of paraquat recovery using H_2O_2 was also examine. Paraquat recovery increased with H_2O_2 content and longer reflux time, but this condition is very dangerous due to the possibility of explosions. Nevertheless, after 1 hour reflux with conc. H_2SO_4 and additional reflux with 30 ml of H_2O_2 for 30 min. increased recovery by 2 ~ 4% . The more effective extraction method is described as follows : Weigh 50 g of soil sample into a 500-ml round-bottomed flask and add 70 ml of conc. HCl, then shake for 1 hr. Filter the HCl portion using Whatman CF/G glass filter, and put the remaining soil sample into a 500 ml round bottomed

Table V . Recovery of paraquat in soils by pre-washing of conc. HCl at different shaking times

Soil	Clay content (%)	Organic matter (%)	Recovery(%)				
				Shaking time for pre-washing (min)			
			control	30	60	120	
A	10.9	1.71	84.0	87.5	96.7	96.8	
B	20.2	0.68	77.0	80.4	84.3	84.9	

flask again, then reflux with 100 ml of H_2SO_4 for 1 hour and then with 30 ml H_2O_2 for 30 min. Adjust the solution to pH 7.0 using 10N-NaOH and add 50 ml of 5 % EDTA-2Na. Readjust the pH to 8.0 to 9.0 and filter. This suggested extraction process requires about 2.5 hours of reflux time (1 hr pre-washing + 1 hr H_2SO_4 + 0.5 hr H_2O_2) and about 300 ml of total volume of filtrate. This relatively small amount of filtrate can reduce the time for cation resin column chromatography compared with 1 L using the ZENECA method. In Table VI, the results of effective extraction procedure are summarized and compared. From the results, it can be suggested that pre-washing with conc. HCl may be helpful in removing the exchange cations of clay minerals such as kaolinite and montmorillonite.

Determination of paraquat

Except for radioimmuno assay and ELISA, spectrophotometry and gas chromatography are more widely used than other instrumental analytical methods. The detection limits of various methods have not improved since 1965, when Calderbank & Yuen[12] reported 0.01 mg/kg detection limits for paraquat using spectrophotometry. Phytotoxicity of paraquat to rice root was observed in some varieties of rice at 10^{-10}~10^{-11} g/l of water concentration. [37,38] This value is much lower than for the wheat bioassay; or for the detection limit of ordinary chemical analysis of 10^{-7} to 10^{-8} g/l level. In Korea, paraquat spray before direct planting of seedlings of rice is usual, and some desorbed paraquat from soils to water can induce a phytotoxicity to the root of transplanted young rice. Therefore, a more sensitive detection method of paraquat residue must be developed. Capillary zone electrophoresis is a newly developed technique that shows good results for this purpose. The electrophoresis technique, which is widely used in protein chemistry, is defined as the transport of electrically charged compounds in solution under the influence of an electrical field. Paraquat ion in paddy water can also be detected by its ionic character. In particular, capillary zone electrophoresis is useful due to the stability over a wide pH range and a long time. Furthermore, low detection amounts of 10^{-13} to 10^{-14} g/l level provide much higher sensitivity and can allow better detection of the paraquat residues in paddy water.

Table VI. Comparison of extraction efficiency among the different methods

Procedure	Korean Standard	Zeneca Method	Suggested by author
Pre-washing	-	-	1 hr, conc. HCl [36]
Reflux time	5 hr. (9 M H_2SO_4)	5 hr. (6 M H_2SO_4)	hr (9M H_2SO_4 1hr + H_2O_2 0.5 hr)
PH adjustment	O	-	O
Addition of EDTA-2Na	O	-	O
Total volume of filtrate	400 ml <	1,000 ml	≈ 300 ml
Possible recovery	70 ~ 80 %	70 ~ 120 %	84 ~ 95 %

References

1. Akhavein, A. A. and Linscott, D. L. (1996) *Residue Rev.* 23 97-145
2. Hance, R. J. (1988) Adsorption and bioavailability, in Environmental Chemistry of Herbicides (R.Grover, ed.) Vol 1. 1-19, CRC Press Inc. Boca Raton, Florida.
3. Hance, R. J. (1967) *J. Sci. Fd Agric.*, 18 544-547
4. Lee, S. J., Katayama, A. and Kimura, M. (1995) *J. Agric. Food Chem.*, 43 1343-1347
5. Tucker, B. U., Pack, C. E., and Ospenson, J. M., (1967) *J. Agric. Food Chem.*, 15 1005-1008
6. Hance, R. J., Byast, T. H. and Smith, P. D. (1980) *Biol. Biochem.* 12 447-448
7. Kanagawa, J. (1990) *Shokacho*, 24, 338 (in Japanese)
8. Imai, Y. and Kuwatsuka, S. (1988) *Pestic. Sci.* 14 475-480
9. Lott, P. F. and Lott, J. W. (1978) *J. Chromatog. Sci.* 16 390-395
10. Ganesan, M., Natesan, S. and Ranganathan, V. (1979) *Analyst*, 104 258-261
11. Yanez, S. P. and Polodiez, L. M. (1986) *Talanta*, 33 745-747
12. Calderbank, A. and Yuen, S. H. (1965) *Analyst*, 90 99-106
13. Yuen, S. H., Bangness, J. E. and Myles. 17, (1967) *Analyst*, 92 375-381
14. Shivahare, P. and Gupta, V. K. (1991) *Analyst*, 116 391-393
15. Kesari, R., Rai, M. and Gupa, V. K. (1997) *Journal of A.O.A.C. International*, 80 388-391

16. Kahn, S. U. (1974) *J. Agr. Food Chem.* 22 863-867
17. Worobey, B. L. (1987) *Pestic. Sci.* 18 245-257
18. Hodgeson, J. W., Bashe W. J. and Eichelberger J. W. (1992) EPA Method 549.1 U.S EPA
19. Kuo, T. L (1987) *Forensic Sci. Int.* 33 177-185
20. Niewol, Z., Hayward, C., Symington, B. A. and Rosan, R. T. (1985) *Clin. Chim. Acta.* 148 149-156
21. Cai, J. and Rassi. Z. E. (1992) *J. Liq. Chromatog.* 15 1193-1200
22. Song, X. and Budde, W. L. (1996) *J. Am. Soc. Mass spectrum* 7 981-986
23. Kwon, J.W. and Lee K.S. (1998) SETAC/UNIDO Asia/Pacific Regional Symposium/Workshop on Environmental Risk Assessment and Management of Chemicals, PC07, 105.
24. Agricultural Chemicals Industrial Association (1997 ~ 1999) Agrochemical Year Book (in Korean).
25. Zeneca Korea Limited (1998) Paraquat documents for special review.
26. Lee, S. G., In, S. H., Chung, Y. H. and Gu, G. S. (1996) *Kor. J. Environ. Toxicol.,* 11: 17-22
27. Agricultural Chemicals Industrial Association (1996) Technical report.
28. Agricultural Chemicals Industrial Association (1998) Technical report.
29. Taylor, T. D. and Amilin, H. I. (1963) *Proc. Southern Weed Conf.,* 16, 405
30. Soil Science Society of America (1997) Glossary of Soil Science Terms, 102.
31. Lee, K. S. *et al.* (1994) Technical report for paraquat dissipation in soil.
32. Agricultural Chemicals Industrial Association : Technical report for paraquat dissipation in soil. (Kim, J.U)
33. Zeneca Agrochemicals (1994) : SOP, RAM 253/01.
34. Zeneca Agrochemicals (1994) : SOP, RAM 254/01.
35. Zeneca Agrochemicals (1995) : SOP, RAM 252/02.
36. Kwon, J. W. and Lee, K. S. (1997) *Korean Journal of Environmental Agriculture,* 16 239- 244.
37. Kwon, Y. W. (1997) personal communication
38. Yoo, J. L. (2001) : Growth inhibition of some rice varieties by paraquat and improvement of analytical method of paraquat in water using capillary electrophoresis, A master thesis of the graduate school of Chungnam National University.

Chapter 12

Air–Soil and Air–Water Exchange of Chiral Pesticides

Terry F. Bidleman[1], Andi D. Leone[2], Renee L. Falconer[3], Tom Harner[1],
Liisa M. Jantunen[1], Karin Wiberg[4], Paul A. Helm[5], Miriam L. Diamond[6],
and Binh Loo[7]

[1]Meteorological Service of Canada, 4905 Dufferin Street, Downsview,
Ontario M3H 5T4, Canada
[2]885 North Hubbard Road, Lowellville, OH 44436
[3]Department of Chemistry, Chatham College, Woodland Road,
Pittsburgh, PA 15232
[4]Department of Chemistry, Environmental Chemistry, Umeå University,
SE–901 87 Umeå, Sweden
[5]Department of Chemical Engineering and Applied Chemistry, University
of Toronto, Toronto, Ontario M5S 3E5, Canada
[6]Department of Geography, University of Toronto, Toronto,
Ontario M5S 3G3, Canada
[7]Hawaii Department of Agriculture, 1428 South King Street, Honolulu, HI 96814

The enantiomers of chiral pesticides are often metabolized at different
rates in soil and water, leading to non-racemic residues. This paper
reviews enantioselective metabolism of organochlorine pesticides (OCPs)
in soil and water, and the use of enantiomers to follow transport and fate
processes. Residues of chiral OCPs and their metabolites are frequently
non-racemic in soil, although exceptions occur in which the OCPs are
racemic. In soils where enantioselective degradation and/or metabolite

formation has taken place, some OCPs usually show the same degradation preference; e.g. depletion of (+)trans-chlordane (TC) and (-)cis-chlordane (TC), and enrichment of the metabolite (+)heptachlor exo-epoxide (HEPX). The selectivity is ambivalent for other chemicals; preferential loss of either (+) or (-)o,p'-DDT and enrichment of either (+) or (-)oxychlordane (OXY) occurs in different soils.

Non-racemic OCPs are found in air samples collected above soil which contains non-racemic residues. The enantiomer profiles of chlordanes in ambient air suggests that most chlordane in northern Alabama air comes from racemic sources (e.g. termiticide emissions), whereas a mixture of racemic and non-racemic (volatilization from soil) sources supplies chlordane to air in the Great Lakes region. Chlordanes and heptachlor exo-epoxide (HEPX) are also non-racemic in arctic air, probably the result of soil emissions from lower latitudes.

The (+) enantiomer of α-hexachlorocyclohexane (α-HCH) is preferentially metabolized in the Arctic Ocean, arctic lakes and watersheds, the North American Great Lakes and the Baltic Sea. In some marine regions (Bering and Chukchi seas, parts of the North Sea) the preference is reversed and (-)α-HCH is depleted. Volatilization from seas and large lakes can be traced by the appearance of non-racemic α-HCH in the air boundary layer above the water. Estimates of microbial degradation rates for α-HCH in the eastern Arctic Ocean and an arctic lake have been made from the enantiomer fractions (EFs) and mass balance in the water column. Apparent pseudo first-order rate constants in the eastern Arctic Ocean are 0.12 y^{-1} for (+)α-HCH, 0.030 y^{-1} for (-)α-HCH, and 0.037 y^{-1} for achiral γ-HCH. These rate constants are 3-10 times greater than those for basic hydrolysis in seawater. Microbial breakdown may compete with advective outflow for long-term removal of HCHs from the Arctic Ocean. Rate constants estimated for the arctic lake are about 3-8 times greater than those in the ocean.

Introduction

Many organochlorine pesticides (OCPs) are global contaminants, dispersed primarily through the atmosphere. Decades after being de-registered in Canada and the United States, OCPs are still routinely found in air and precipitation from the Great Lakes region *(1-5)* and the Arctic *(6-8)*. International treaties and regional action plans have been established to ban or phase out persistent organic pollutants (POPs) *(9,10)*. Twelve POPs, eight of which are OCPs (aldrin, dieldrin, DDT, hexachlorobenzene, chlordane, heptachlor, mirex and toxaphene), were recently banned under the United Nations Environmental Program (UNEP) Global POPs Protocol *(9,10)*. Although not on the UNEP list, technical hexachlorocyclohexane (HCH) is included among the POPs targeted under the United Nations Economic Commission for Europe Convention on Long-Range Transport (UN-ECE-LRTAP) *(9)*.

A new approach to investigating pesticide transport and fate in the environment uses chiral analytical methods to determine individual enantiomers. About one-fourth of the agrochemicals used worldwide today are chiral *(11)*. Most are produced as racemates, although the trend toward single-enantiomer chemicals is growing rapidly. There are several advantages of using single-enantiomer pesticides: lower production cost, smaller quantities needed for effective pest control, and less environmental damage *(12)*. Several classes of pesticides have members that are chiral: organochlorines, organophosphates, pyrethroids and herbicides. Chiral OCPs include o,p'-DDT, α-HCH, heptachlor (HEPT), several components of technical chlordane and technical toxaphene, and the metabolites oxychlordane (OXY), heptachlor exo-epoxide (HEPX) and o,p'-DDD (Figure 1).

Although a few modern chiral pesticides are now manufactured as single-enantiomer products, most are racemates. Enantiomers have the same physical and chemical properties, therefore abiotic transport processes (leaching, volatilization, atmospheric deposition) and reactions (hydrolysis, photolysis) are not likely to be enantioselective (some exceptions are discussed in the Conclusions section). Metabolism by microorganisms in water and soil often proceeds enantioselectively, leading to non-racemic residues.

In this article we review applications of pesticide enantiomers for tracing transport and transformation processes. We specifically focus on air-soil and air-water exchanges and selective degradation pathways in the physical environment. Details of our own studies and analytical methods are given in primary publications *(13-23)*. Chiral methods have also been applied in several recent food web and bioaccumulation studies *(24-32 and references cited therein)*, although these are not discussed here. Reviews of chiral analytical methods and enantioselective metabolism *(12,33,34)* of xenobiotics and a book devoted to these subjects *(35)* have been published.

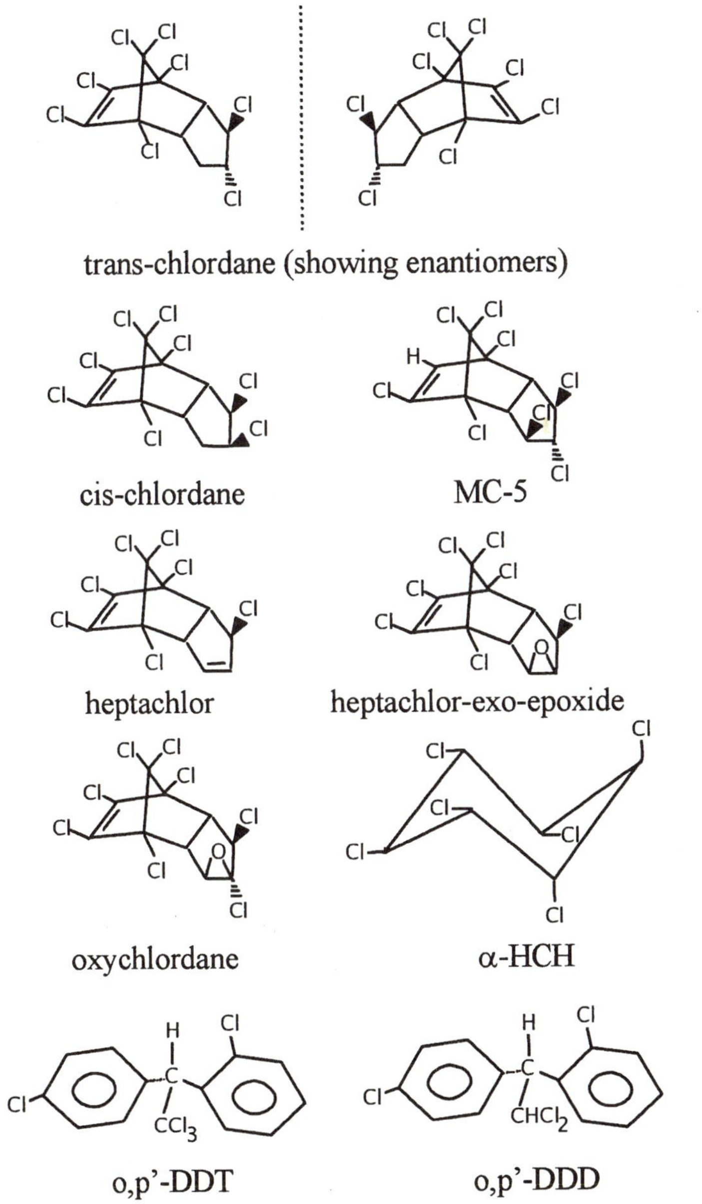

trans-chlordane (showing enantiomers)

cis-chlordane

MC-5

heptachlor

heptachlor-exo-epoxide

oxychlordane

α-HCH

o,p'-DDT

o,p'-DDD

Figure 1. Structures of chiral organochlorines and their metabolites.

Enantioselective Metabolism of Organochlorine Pesticides in Soils

Pesticide Residue Concentrations

Quantitative measurements of OCP residues in soils are useful for several purposes: assessing the reservior of pesticides held in the terrestrial environment, determining bioaccumulation in plants and terrestrial biota (e.g., earthworms) and modeling soil-air exchange. Chiral Ocps in soils have been measured at several locations in North America. Residues have been determined in agricultural soils of Alabama *(13,14)*, the midwestern U.S. (Ohio, Indiana, Illinois and western Pennsylvania) *(15,16)* and the Fraser Valley of British Columbia, Canada *(17,18)*. Compounds sought were o,p'-DDT (and other achiral DDTs), chlordanes, HEPT, OXY, HEPX, α-HCH and, in Alabama, toxaphene (though no chiral studies were done for toxaphene). In Connecticut, chlordanes only were determined in agricultural, home lawn and garden, and house foundation soils *(36-39)*. Results from ten soil samples collected from farms on Molokai, Hawaii and analyzed using previously published methods *(15-17)* are reported here.

Residue data from these studies are summarized in Table I as ranges, arithmetic means (AM) or geometric means (GM) depending on the information given in the original publication. Although there have been other quantitative determinations of OCPs in soils, only the locations where enantiomeric measurements have also been made are included in Table I. Concentrations in the soils varied by several orders of magnitude, even within a relatively small region. Log normal distributions and GM are probably the best measures of variability and central value *(14,15)*, however some reports do not give GM nor the individual sample data to calculate them. OCP residues did not correlate with the percent organic carbon for most soils in Alabama (0.3-3.5% organic carbon) *(14)*, nor in the midwest U.S. (0.7-7.6% organic carbon) *(15)*, however one midwest soil with 33% organic carbon contained exceptionally high levels of OCPs. In British Columbia, OCPs concentrations increased from loamy sand (1.0-1.8% organic matter), to silt loams (3.7-6.5% organic matter) to muck soils (27-56% organic matter) *(40)*.

Ratios of p,p'-DDT to the metabolite p,p'-DDE (DDT/DDE) are often used to roughly estimate the age of the residues, since the proportion of p,p'-DDE tends to increase with time. This ratio in the individual soil samples varied from 0.5 - 1.4 in Hawaii, 0.5 - 1.5 in Alabama *(14)*, and 0.5 - 6.6 in the midwest U.S. *(15)*. DDT/DDE ratios derived from GM concentrations were Hawaii = 0.88, Alabama = 0.84, midwest U.S. = 1.25. Chiral OCPs were reported in soil samples from the U.K. which had been archived between 1968-90 *(41)*. The samples came from experimental plots; one was amended with sewage sludge in 1968 and the other was an unamended control site. Residues of hexachlorobenzene, DDE and dieldrin were slightly higher and significantly different (p >0.05) in the sludged plot compared to the control plot. Ratios of DDT/DDE varied between 0.14 - 1.0, but showed no trend with time.

Chlordane residues in soils representing three land use classes in Connecticut were determined: agricultural, home lawn and garden and house foundation *(36,38)*. Total chlordane (trans-chlordane + cis-chlordane + trans-nonachlor = TC + CC + TN) was highest in foundation soils (mean = 7180 ng/g dry wt.), followed by residential (330 ng/g) and agricultural (142 ng/g) soils. A detailed study was also carried out on one lawn site located at the Connecticut Agricultural Experiment Station where chlordane had been applied once in 1960 and had remained under continuous grass cover until 1998, when the soil samples were collected prior to tilling the site for planting crops. Total chlordane averaged 7689 ng/g dry wt. Within the treatment site, the chlordane components were in the proportion CC = TC > TN, and this order was also found for house foundation soils where residues exceeded 5000 ng/g dry wt. Chlordane residues in soil samples taken ~2-40 m outside the treated site were approximately 1% of those inside, and proportions of the three chlordane components showed considerable variation. The investigators also showed that weathered chlordane residues from the experimental site were accumulated in several crops via root uptake and deposition of atmospheric residues onto leaves *(37,39)*.

Trans-chlordane is thought to be metabolized more rapidly than cis-chlordane in soils *(38)*, and this should be indicated by the ratio of TC/CC. Differences in the proportions of chlordane isomers are found among soils, as noted above for sites in Connecticut. The ratios of TC/CC in the U.K. soils ranged from 0.6 - 2.9 *(41)*. TC/CC ratios in the individual Hawaiian soils ranged from 4.1 - 6.5, and averaged 5.3 when calculated from GM concentrations. Similar statistics were 0.54 - 1.7 (mean = 0.94) in Alabama. *(14)*, and 0.33 - 6.0 (mean = 1.1) in the midwest U.S. *(15)*. The TC/CC ratio in technical chlordane is 0.95-1.18 *(36,42)*. Overall, Hawaiian soils contained a greater proportion of trans-chlordane than soils from the other regions. This might be due to the use of heptachlor in Hawaiian agriculture, since technical heptachlor is contaminated with chlordane isomers *(43)*.

Enantioselective Processes

Organochlorine Pesticides

Enantioselective metabolism of chiral pesticides often takes place in soils, leading to the presence of non-racemic residues. Resolution of the individual

Table I. Chiral Organochlorine Pesticides in Soils, ng/g Dry Wt.

Location	Year	N^a	Range or AM^b	S.D.	GM^b	Ref.
DDT[c]						
Alabama	1996	38	54	38	17	*14*
Midwest U.S.	1995-96	34	<0.5-11800		9.6	*15*
Hawaii	1999	10	1540	1200	1080	*This work*
Brit. Columbia: muck	1989	4	2984-7162			*40*
Brit. Columbia: silt loam	1989	4	194-1275			*40*
U.K.[d]	1968-90	6	1.4-3.2			*41*
U.K.[e]	1968-90	6	2.0-9.2			*41*
Chlordane[f]						
Alabama	1996	32	1.9	1.1	0.6	*14*
Midwest U.S.	1995-96	32	<0.2-847		1.3	*15*
Hawaii	1999	10	110	71	88	*This work*
Brit. Columbia: muck	1989	4	284-1943			*40*
Brit. Columbia: silt loam	1989	4	0.6-338			*40*
U.K.[d]	1968-90	6	0.17-0.54			*41*
U.K.[e]	1968-90	6	0.09-0.81			*41*
Connecticut: exp. site	1997-99	35	7689	4340		*37,38*
Connecticut: agric.	1997-99	53	142	120		*37,38*
Connecticut: lawn/garden	1997-99	34	330	987		*37,38*
Connecticut: house found.	1997-99	45	7180	14296		*37,38*
Heptachlor						
Alabama	1996	3	0.04	0.02	0.04	*14*
Midwest U.S.	1995-96	16	<0.04-56		0.11	*15*
Hawaii	1999	10	9.4	8.7	2.7	*This work*
Brit. Columbia: muck	1989	4	37-278			*40*
Brit. Columbia: silt loam	1989	4	0.16-1.0			*40*

Compound	Location	Year	n[a]	Range	AM[b]	GM[b]	Ref.
HEPX[g]	Alabama	1996	12	0.19	0.14	0.10	14
	Midwest U.S.	1995-96	27	<0.04-121		0.49	15
	Hawaii	1999	10	111	77	86	This work
	Brit. Columbia: muck	1989	4	85-390			40
	Brit. Columbia: silt loam	1989	4	0.05-36			40
α-HCH	Alabama	1996	24	0.10	0.12	0.05	14
	Midwest U.S.	1995-96	12	<0.04-1.2		0.09	15
	Brit. Columbia: muck	1989	4	10-259			40
	Brit. Columbia: silt loam	1989	4	0.26-2.3			40
	U.K.[d]	1968-90	6	0.04-0.37			41
	U.K.[e]	1968-90	6	0.06-0.24			41
Toxaphene	Alabama	1996	39	285	390	84	14

a) Number of samples.
b) AM = arithmetic mean, GM = geometric mean.
c) Some or all of: p,p'-DDT, o,p'-DDT, p,p'-DDE, o,p'-DDE, p,p'-DDD, o,p'-DDD.
d) Archived soils from one control (unsludged) site, only the range is given because the samples were from different years.
e) Archived soils from one site, treated once with sewage sludge in 1968.
f) Some or all of cis-chlordane, trans-chlordane, trans-nonachlor.
g) HEPX = heptachlor exo-epoxide, includes oxychlordane for midwest U.S. soils.

enantiomers of OCPs can be accomplished by capillary gas chromatography on derivatized cyclodextrin columns with detection by mass spectrometry *(15-17,34,35)*. Results are usually presented as the enantiomer ratio (ER), defined as the concentration of (+)/(-) enantiomers, or the first/second eluting enantiomer when the optical signs are not known *(33-35)*. Here we use the enantiomer fraction, EF = (+)/[(+) + (-)], which is preferred over ER *(44,45)*.

EF values for chiral OCPs and metabolites in soils of different regions are given in Table II. In our work, replicate injection of racemic standards (EF = 0.50) yielded standard deviations of 0.015 or less *(15-17)*. For parent OCPs, EF values <0.50 and >0.50 indicate preferential metabolism of the (+) and (-) enantiomer, respectively. In the case of metabolites OXY and HEPX, EF >0.50 or <0.50 could arise from preferential formation of the (+) or (-) product from the parent compound, and/or selective degradation of the opposite enantiomer of the metabolite. The EF values in Table II suggest regional differences in the metabolism of OCPs, although only a few soils from each region have been analyzed by chiral methods.

Enantioselective metabolism in most agricultural soils favored degradation of (+)TC and (-)CC. Exceptions were British Columbia, where racemic chlordanes were found in the six soils examined *(17)*, and at one farm in Connecticut *(38)*. Two other Connecticut farms, and a test plot at the Connecticut Agricultural Experiment Station, contained non-racemic chlordanes. Soils near house foundations in Connecticut contained very high chlordane residues which were racemic. The mean EF of TC in midwest U.S. agricultural soils (0.41) *(15)* was significantly different ($p < 0.01$) from the mean EFs in Alabama (0.48) *(23)* and Hawaii (0.46) *(this work)*. Differences in EFs of CC among the three regions (0.52 - 0.55) were not significant ($p > 0.05$). Eitzer et al. *(38)* found a strong (negative) correlation between the ERs of TC and CC in Connecticut soils, and demonstrated that a similar correlation existed in the data of Aigner et al. *(15)* for midwest U.S. soils.

Technical chlordane was applied to the test plot in Connecticut once in 1960 and the plot lay under grass cover until it was tilled in 1998. Investigation of the site before and after tilling showed no significant difference in total chlordane residues, but significant differences ($p < 0.05$) were found in the ERs of TC and CC, and in the ratios TC/TN and CC/TN. Soil tillage appeared to enhance enantioselective degradation *(38)*. The U.K. experimental plots showed significantly ($p < 0.05$) non-racemic EFs for both TC (EF = 0.40 - 0.46) and CC (EF = 0.52 - 0.57), although there was no significant difference between sludged and unsludged soils nor a trend with time from 1968-90 *(41)*.

In some studies EFs of the chlordane component MC5 (Figure 1) were also determined. The elution order of (+) and (-) enantiomers are not known for

MC5, so the EF was calculated as A1/(A1 + A2), where A1 and A2 refer to the areas of peaks eluting first and second on a Betadex-120 or Gammadex-120 chiral phase column *(15-17,23,38)*. In soils showing enantioselective degradation, the first-eluting enantiomer of MC5 was depleted (EF <0.50).

Soils show a strong preference for (+)HEPX accumulation. The EFs in Alabama (0.67) *(23)* and the midwest U.S. (0.71) *(15)* were different from the EF in Hawaii (0.55) (p <0.01), and the EFs in the midwest U.S. and British Columbia (0.56 - 0.58) *(17)* were different from each other (p <0.01). EFs in Alabama and British Columbia were not significantly different (p >0.05) due to the small number of samples in each set (Table II).

Residues of α-HCH were high enough for chiral analysis in British Columbia *(17)*, the U.K. *(41)*, a few soils in Alabama *(23)*, and one midwestern soil *(15)*. The α-HCH in three British Columbia silt loams was racemic (EF range = 0.49 - 0.50), whereas selective breakdown of (-)α-HCH was found in three muck soils (EF range = 0.55 - 0.58). The same selectivity was found for one soil sample near a former HCH factory in Germany (EF = 0.52) *(44)*. The EF averaged 0.50 in seven Alabama soils and was 0.49 in the one midwest soil. EFs in the U.K. soils ranged from 0.48 - 0.52 and were generally not significantly different (p >0.05) from racemic in either sludged or unsludged treatments.

Residues of OXY and o,p'-DDT showed selective accumulation of the (+) enantiomer in some soils and the (-) enantiomer in others (Table II). In the case of OXY, the EFs in soil may be the net result of selective metabolism of (+)TC and (-)CC to OXY enantiomers with these respective signs, as suggested for biota *(26)*. Eitzer et al. *(38)* found a positive correlation (r^2 = 0.68) between the CC/TC ratio and percent OXY in Connecticut soils, demonstrating more rapid conversion of TC rather than CC to OXY. Reasons for the ambivalence of o,p'-DDT are not known.

Soils from Alabama *(14)* and other southern U.S. states *(47)* contain residues of toxaphene resulting from former usage on cotton, soybeans and other crops. So far, no chiral studies have been done on these residues, but the chromatographic pattern in soil *(47)* and ambient air *(42)* indicates selective depletion of certain labile congeners, notably B8-531(2,2,3-exo,5-endo,6-exo,8,9,10-octachlorobornane) and B8-806/809 (2,2,5-endo,6-exo,8,9,10 and 2,2,5-endo,6-exo,8,9,9,10-octachlorobornanes). Vetter and Maruya *(48)* found enantioselective depletion of certain toxaphene congeners in the lower food web from a toxaphene-contaminated salt marsh in Georgia, although the ones in soils mentioned above were not investigated.

Modern Pesticides

There have been several recent studies of the enantioselective degradation of modern pesticides in soil and water. A few examples are briefly mentioned

Table II. Enantiomer Fractions (EF) of Organochlorine Pesticides in Soils[a]

	Location	Year	N[b]	Range	AM[c]	S.D.	Ref.
o,p'-DDT	Alabama	1996	30	0.41-0.57	0.50	0.036	23
	Midwest U.S.	1995-96	16	0.43-0.54	0.49	0.038	15
	Hawaii	1999	9	0.45-0.50	0.48	0.014	This work
	Brit. Columbia: muck	1989	3	0.50-0.51	0.51	0.004	17
	Brit. Columbia: silt loam	1989	3	0.45-0.51	0.49	0.029	17
	U.K.[d]	1968-90	10	0.48-0.57	0.51	0.027	41
Trans-chlordane	Alabama	1996	11	0.46-0.50	0.48	0.012	23
	Midwest U.S.	1995-96	23	0.32-0.48	0.41	0.040	15
	Hawaii	1999	10	0.44-0.49	0.46	0.018	This work
	Brit. Columbia: muck	1989	3	0.49-0.50	0.49	0.001	17
	Brit. Columbia: silt loam	1989	3	0.49-0.50	0.50	0.003	41
	U.K.[d]	1968-90	11	0.40-0.46	0.43	0.019	41
	Connecticut: exp. site	1997-99	35		0.46	0.02	38
	Connecticut: agric.	1997-99	13	0.41-0.50	0.47	0.02	38
	Connecticut: house found.	1997-99	15		0.49	0.01	38
Cis-Chlordane	Alabama	1996	11	0.48-0.56	0.53	0.022	23
	Midwest U.S.	1995-96	23	0.50-0.61	0.55	0.029	15
	Hawaii	1999	10	0.49-0.53	0.52	0.011	This work
	Brit. Columbia: muck	1989	3	0.50	0.50	0.000	17
	Brit. Columbia: silt loam	1989	3	0.49-0.50	0.50	0.006	17
	U.K.[d]	1968-90	11	0.52-0.57	0.54	0.017	41

				EF	AM		Ref.
HEPX[e]	Connecticut: exp. site	1997-99	35		0.54	0.03	38
	Connecticut: agric.	1997-99	13	0.51-0.55	0.53	0.01	38
	Connecticut: house found.	1997-99	15		0.50	0.005	38
	Alabama	1996	4	0.50-0.76	0.67	0.12	23
	Midwest U.S.	1995-96	14	0.54-0.88	0.71	0.087	15
	Hawaii	1999	10	0.53-0.57	0.55	0.013	This work
	Brit. Columbia: muck	1989	3	0.55-0.58	0.56	0.016	17
	Brit. Columbia: silt loam	1989	1	0.58	0.58		17
Oxychlordane	Alabama	1996	3	0.55-0.60	0.54	0.023	23
	Midwest U.S.	1995-96	17	0.40-0.62	0.52	0.066	15
	Hawaii	1999	9	0.52-0.65	0.60	0.049	This work
	Brit. Columbia: muck	1989	3	0.37-0.46	0.42	0.047	17
	Brit. Columbia: silt loam	1989	1	0.40	0.40		17
α-HCH	Alabama	1996	7	0.48-0.53	0.50	0.013	23
	Midwest U.S.	1995-96	1	0.49	0.49		15
	Brit. Columbia: muck	1989	3	0.55-0.58	0.56	0.015	40
	Brit. Columbia: silt loam	1989	3	0.49-0.50	0.50	0.007	40
	U.K.[d]	1968-90	11	0.48-0.52	0.50	0.010	41
	Germany		1	0.52	0.52		46

a) EF = (+)/[(+) + (−)] enantiomers.
b) Number of samples.
c) AM = arithmetic mean.
d) Archived soils, sludge-treated once in 1968 and control, followed from 1968-90.
e) HEPX = heptachlor exo-epoxide.

below and in the next section, and for the phenoxy herbicide mecoprop, in a recent review *(12)*.

Enantioselective degradation of the phenoxy herbicide dichlorprop after field application of the racemic mixture was followed by capillary electrophoresis using the chiral reagent heptakis(2,3,6-tri-O-methyl)-β-cyclodextrin *(49)*. Breakdown of the herbicide occurred almost completely within a month. The S(-) enantiomer degraded significantly faster than the R(+) enantiomer, in contrast to the results of a laboratory study with marine microorganisms which degraded the R(+) form exclusively.

The racemic fungicide metalaxyl is currently being replaced by metalaxyl-M, which is enriched with the active R(-) enantiomer. Degradation of the two enantiomers of metalaxyl and its initial breakdown product MX-acid in experimental soil cultures proceeded enantioselectively, with the R(-) enantiomer being converted to its corresponding MX-acid faster than S(+) metalaxyl. No enantiomerization (interconversion of the two enantiomers) was noted during the degradation for metalaxyl or MX-acid *(50)*.

Lewis et al. *(51)* studied the effects of simulated global warming, deforestation and nutrient addition on the enantioselective degradation in soil of the organophosphorus insecticide ruelene and the herbicides dichlorprop and methyl dichlorprop. Methyl dichlorprop was demethylated quickly ($t_{1/2}$ ~0.7 d), whereas degradation of parent dichlorprop and ruelene proceeded only after lag times of ~6 months. Simulated global warming was investigated by raising the temperature of North American and Norwegian soils by 5 °C over a 4-7 year period. Soil samples were then collected and incubated with pesticides in the laboratory. Neither heating, nor addition of inorganic nutrients, affected the enantioselectivity for methyl dichlorprop demethylation. Addition of organic nutrients (beef broth) caused soils which preferentially demethylated the (+) enantiomer to strongly prefer the (-) enantiomer in soils from North America and Brazil, but not in Norway where soil degradation already favored the (-) enantiomer. In Brazil, deforestation caused soils to shift from preferentially removing the (-) enantiomer of ruelene to exclusively degrading the (+) enantiomer. Soil warming in Norway caused soils to shift from all samples degrading (+)ruelene to 22% preferring (-)ruelene.

Lewis et al. *(51)* suggested that microbial enantioselectivity towards environmental pollutants is controlled by the activation of metabolically quiescent microbial populations or the induction of enantiomer-specific enzymes. Changes in enantioselectivity could be due to different groups of related microbial genotypes being activitated under different environmental conditions. This is also suggested by the increased enantioselective degradation of TC and CC after tilling the soil at the Connecticut test plot *(38)*.

Reductive dechlorination of o,p'-DDT was investigated in the laboratory using the aquatic plant elodea (*Elodea canadensis*) and the terrestrial plant kudzu (*Pueraria thunbergiana*), a partially purified fraction of an *Elodia* extract, the porphyrin hematin and human hemoglobin *(51)*. Although all of these systems degraded o,p'-DDT with formation of o,p'-DDD, none was enantioselective.

Enantioselective Metabolism of Pesticides in Water

Organochlorine Pesticides in Marine Waters

Most investigations of chiral pesticides in the aquatic environment have been for α-HCH. EFs in these studies are summarized in Table III. Early work showed enantioselective degradation of α-HCH in the North Sea *(53-55)*. The degradation preference was ambivalent, with some regions of the North Sea showing depletion of the (+) enantiomer, and the (-) enantiomer in others *(53)*. Enantioselective degradation of α-HCH is accompanied by formation of two chiral β-pentachlorocyclohexenes (β-PCCHs), whereas degradation of achiral γ-HCH produces two chiral γ-PCCHs. The relative proportions of these isomers and their respective enantiomers in the aquatic environment are controlled by rates of formation by dehydrochlorination of the HCHs and degradation of the PCCHs *(54)*. Investigations in Skagerrak *(54)* and the southern Baltic Sea *(22,54)* showed uniform depletion of (+)α-HCH. Selective degradation of (-)α-HCH was found in the Bering-Chukchi seas *(19)* but (+)α-HCH was degraded preferentially in the western *(19,20,28,56)* and eastern *(21,57)* Arctic Ocean. The α-HCH residues were slightly non-racemic in surface water of the South Atlantic and Southern Ocean between South Africa and Antarctica, with differences in enantiomer preference among regions *(58)*.

When racemic α-HCH was added to a culture of microorganisms isolated from aerobic marine sediment, (+)α-HCH was preferentially dehydrochlorinated to form non-racemic β-PCCH *(54,59)*. Anaerobic breakdown of α-HCH in active sewage sludge also favored the (+) enantiomer. A slower rate of degradation was found in sterilized sludge with no chiral preference *(60)*.

The EF of α-HCH decreases with depth in the western *(19)* and eastern *(21,57)* Arctic Ocean, due to preferential loss of (+)α-HCH in the deeper, older water layers (Figure 2). Degradation rates of γ-HCH and the two enantiomers of α-HCH in the eastern Arctic Ocean were estimated *in situ* from measured vertical profiles of concentration and enantiomer composition (for α-HCH) in

210

Table III. Enantioselective Degradation of α-HCH in Oceans and Lakes[a]

Location	Year	Range of EF	Ref.
Marine			
Bering Sea	1993	0.52-0.53	*19*
Chukchi Sea	1993	0.51-0.55	*19*
Western Arctic Ocean			
Surface - 50 m	1992-94	0.41-0.48	*19,20,56*
Surface	1998	0.45	*28*
300 - 2200 m	1994	0.11-0.44	*19*
Eastern Arctic Ocean			
Surface - 50 m	1996	0.42-0.48	*21,57*
200 - 1000 m	1996	0.11-0.37	*21,57*
North Sea	1987	0.44-0.54	*53-55*
	1994-95	0.45-0.48	*73*
Baltic Sea	1987	0.44-0.48	*54*
	1997-98	0.44-0.46	*22*
South Atlantic Ocean	1997	0.49-0.52	*58*
Southern Ocean	1997	0.48-0.51	*58*
Freshwater - Canadian Arctic and Subarctic			
Lakes, Eastern Arctic	1992-98	0.36-0.43	*56,61,62,64*
Lakes, Yukon	1993	0.24-0.47	*63,64*
Streams and wetlands (Eastern Arctic)	1992-98	0.26-0.50	*56,61,62,64*
Freshwater - Canadian Temperate			
Small Ontario lakes	1999	0.47-0.51	*64*
Ontario streams and wetlands	1999	0.39-0.47	*64*
Great Lakes	1993-98	0.46-0.47	*13,64*

a) Surface water, unless stated otherwise.

vertical profiles of concentration and enantiomer composition (for α-HCH) in the water column between 200-800 m, and the estimated total loss of HCHs from that water mass between 1979-96 *(21,57)*. Pseudo first-order microbial degradation rates were 0.12 y^{-1} for (+)α-HCH, 0.030 y^{-1} for (-)α-HCH and 0.037 y^{-1} for γ-HCH *(21)*. These were 3-10 times faster than basic hydrolysis rates in seawater. As a result, microbial degradation may account for a third of the total annual HCH loss from the Arctic Ocean, competing with advective outflow *(57)*.

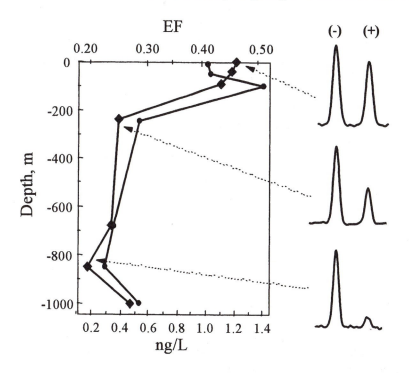

Figure 2. Enantiomer chromatograms (right) and vertical profiles of concentration (circles) and EF (diamonds) with depth of α-HCH in the Eastern Arctic Ocean. Note the preferential loss of (+)α-HCH in the deeper, older water layers. (Adapted with permission from ref. 57. Copyright American Geophysical Union).

The enantiomer composition of chlordane and HEPX in Arctic Ocean surface water was determined in a 1994 study (10 - 14 samples) *(20)*. TC and CC were both racemic (EF = 0.50 ± 0.006 and 0.50 ± 0.008, respectively). As in soils, HEPX was enriched in the (+) enantiomer (EF = 0.62 ± 0.011).

Fresh Waters

Organochlorine Pesticides

The (+) enantiomer of α-HCH was selectively depleted in Canadian arctic lakes, streams and wetlands *(56,61-64)* and in the North American Great Lakes *(13,64)*. Factors influencing the enantioselective degradation in lakes and wetlands have been investigated *(62-64)*. Preferential degradation of (+)α-HCH was extensive in arctic watersheds and wetlands resulting in EF = 0.27 - 0.47 *(61-64)*. Arctic lakes also exhibited low EF = 0.24 - 0.47 *(56,61-64)*. By contrast, α-HCH in small lakes from temperate regions of Canada was more nearly racemic (EF = 0.47 - 0.51), and the range of temperate wetland and stream EFs (0.39 - 0.47) indicated somewhat less enantioselectivity than in arctic wetlands *(64)*. Law et al. *(64)* put forward the hypothesis that enantioselective degradation is optimized by maximal contact between the α-HCH and sediment substrates in nutrient-poor water where oligotrophic bacteria might act as biofilms. Preferential degradation of (-)α-HCH was found in groundwater samples from Florida in the vicinity of a former technical HCH formulations plant *(65)*.

Comparisons were made of α-HCH degradation in one arctic lake (Amituk Lake) and its inflowing streams *(62)*. Approximately 7% of the α-HCH was enantioselectively degraded in the watershed prior to entering the lake. A mass balance in the lake showed that over-winter loss amounted to 34 - 63% of the water column inventory. Sedimentation was negligible, outflow in winter was zero and hydrolysis was estimated to account for only 2% of the loss. The remaining loss was attributed to microbial degradation, with an apparent pseudo first-order rate constant of 0.48 - 1.13 y^{-1}. This value is about 3 - 8 times greater than the combined microbial degradation rate for the two α-HCH enantiomers in the Arctic Ocean *(21)*. Most of the over-winter degradation within the lake appeared to be due to non-enantioselective processes.

Modern Pesticides

Studies have been carried out on the enantioselective degradation and interconversion of enantiomers of the herbicides mecoprop and dichlorprop in water *(66,67)*. Only the R(+)enantiomers are herbicidally active, and these were degraded more slowly than the inactive S(-)enantiomers *(66,67)*. In Switzerland, only R(+)mecoprop is registered as a herbicide, but interconversion of enantiomers in the environment led to the S(-)enantiomer being detected in Swiss lakes *(68)*. R(+) and S(-)mecoprop in some lakes *(68)* and a polluted aquifer *(69)* may have also resulted from inputs of racemic mecoprop. The source of this racemic mecoprop was traced to its use in a bituminous roof

against root penetration. The sealant released (R,S)mecoprop when hydrolyzed, which subsequently led to its transport into the environment *(70,71)*.

Metolachlor is a herbicide which contains four stereoisomers due to two chiral elements in its structure, an asymmetric carbon and a chiral axis. Metolachlor was first used as a racemate, but since 1997 has been partially replaced by S-enriched metolachlor which contains about 90% 1'S-isomers *(72)*. Surface water from Swiss lakes sampled prior to 1998 contained racemic metolachlor, but a clear excess of the 1'S-isomers was found in 1998 and 1999, indicating contributions of the newly released, non-racemic herbicide *(72)*. Concentrations and enantiomer compositions in surface and deep waters were followed over two years and a computer simulation was used to estimate that 55% (1998) and 90% (1999) of the racemic metolachlor had been substituted by S-metolachlor. The authors pointed out that there was no marked difference in metolachlor concentrations over this period, thus the switch of metolachlor products could not have been deduced by concentration measurements.

Bromcyclen is a bicycloheptene containing six chlorines and one bromine, used against ectoparasites in Europe. Water samples taken in the River Stör, a tributary of the Elbe in Germany, contained racemic bromcyclen. The bromcyclen in fish from the Stör was non-racemic and differed between trout, which showed depletion of the (-) enantiomer, and bream, in which depletion of the (+) enantiomer was found *(73)*.

Atmospheric Exchange of Chiral Pesticides with Soil and Water

Air-Soil Exchange

The enantiomer signatures of non-racemic pesticides in soil and water are preserved upon volatilization, providing useful tracers for following emissions. Pesticides in air are subject to only abiotic removal processes which are non-enantioselective, such as photolysis, hydroxyl radical attack and atmospheric deposition. Enantiomers have the same physicochemical properties and thus EFs are not subject to fractionation effects as are isomer- and parent/metabolite ratios. For these reasons, EF values in air are conservative and can carry information about the sources of pesticides *(74,75)*.

Volatilization of α-HCH and HEPX into the air above an agricultural field was investigated at a farm in the Fraser Valley, British Columbia *(18)* where (-)α-HCH was depleted and (+)HEPX was enriched in the soil. Figure 3 shows vertical profiles of OCP concentration and EFs in the air above the soil. Although concentrations decline with height for both compounds, there is a distinct difference in the behavior of the EFs. Those for α-HCH also decrease with height, while the HEPX EFs are relatively invariant. For α-HCH, the non-racemic emissions from the soil may be diluted with more nearly racemic α-HCH from distant transport. Background α-HCH in the atmosphere has an EF close to 0.50 *(13,16)*. In the case of HEPX, the soil of this particular field may be the main source to the overlying air, or the background air has a similar EF value. Ambient air samples from the southern U.S. and the Great Lakes region contained non-racemic HEPX, with EF values ranging from 0.60 - 0.67 *(74,76,77)*.

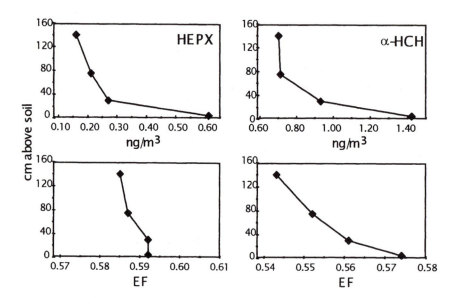

Figure 3. Profiles of OCP concentration (upper) and EF (lower) in air above an agricultural field in the Fraser Valley, British Columbia. Data from (18).

Chlordane was used in the U.S. until 1978 for agriculture, ornamentals, home lawns and gardens, and termite control. After 1983, the only permitted use was as a termiticide, and all uses were cancelled in 1988 *(78)*. The chlordane in ambient air today may come from volatilization of residues in soils, emission of termiticides from treated buildings and long-range transport from countries where chlordane is still used.

Ambient air samples from the Great Lakes region contained non-racemic TC and CC *(16,74,75,77)*, whereas nearly racemic chlordanes were found in the air of Alabama *(42)* and the city of Columbia, South Carolina *(42,74,75)*. Residues of TC and CC in agricultural soils of Alabama *(23)*, the midwest U.S.

air of Alabama *(42)* and the city of Columbia, South Carolina *(42,74,75)*. Residues of TC and CC in agricultural soils of Alabama *(23)*, the midwest U.S. *(15)* and one farm in the U.K. *(41)* were non-racemic (Table II). Indoor air concentrations of chlordanes in Alabama and South Carolina *(42)* and the midwest U.S. *(79)* greatly exceeded those in ambient air, and the TC and CC were racemic. Figure 4 shows the chromatographic profile of non-racemic chlordanes in an Ohio soil. EF values for this soil, indoor air and Great Lakes air are also shown.

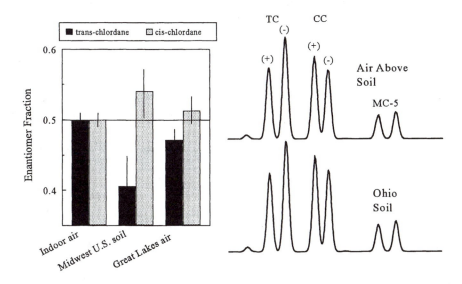

Figure 4. Right: Chromatographic separation of chlordane enantiomers (TC = trans-chlordane, CC = cis-chlordane, and chlordane MC-5) in an Ohio soil sample and overlying air(15,16). Left: EF values for indoor air (79), midwest U.S. soil (15) and Great Lakes air (74,77). Vertical lines show standard deviations. (Adapted from ref. 75, copyright American Chemical Society).

Studies at six farms in the midwestern U.S. were carried out by collecting paired soil and overlying air samples *(16)*. Both soil and air contained non-racemic chlordanes, HEPX, OXY and o,p'-DDT. Vertical profiles of chlordane concentrations and EFs were obtained at one farm. Although concentrations decreased from 10 – 175 cm above the soil, indicating soil-to-air exchange, the EFs were invariant as seen for HEPX in British Columbia (Figure 3) *(18)*. EFs of chlordanes in air-above-soil samples at the six midwestern farms averaged 0.423 ± 0.024 for TC and 0.526 ± 0.017 for CC. Means for 13 ambient air were 0.481 ± 0.005 for TC and 0.509 ± 0.005 for CC. These are intermediate of air-above-soil EFs and the racemic EFs found for indoor air *(75)*.

Mattina et al. *(39)* reported a study of chlordane isomers and enantiomers transferred from soil to vegetable crops and overlying air at the Connecticut Agricultural Experiment Station. Zucchini showed a $2.5 - 8$ fold enhancement of chlordane and nonachlor isomers in the root relative to the soil. Changes in the relative concentrations of chlordane enantiomers were found among bulk soil and various compartments of plants, indicating that enantioselective processes were in effect. EFs of TC (0.46) and CC (0.53) in air at 0.6 m height were most similar to those in soil (0.45 and 0.55 respectively for TC and CC) and approached background air EFs (0.48 and 0.52) with increasing height above the soil. The EFs in leaves and fruit peel of cucumber growing within the 0.6-m zone ranged from $0.30 - 0.37$ for TC and $0.48 - 0.52$ for CC, suggesting that enantioselective processes within the plant, rather than air-plant exchange dominated the observed enantiomer signatures.

Results of these investigations suggest that both racemic and non-racemic sources contribute to observed chlordane EFs in Great Lakes air *(16,74,75,77)*. In contrast, racemic sources appear to dominate chlordane inputs to urban air in South Carolina and ambient air in Alabama *(42,74)*.

Background air in the Northern Hemisphere is likely to be influenced by long-range transport of pesticide residues emitted from contaminated soils *(80)*. Air samples collected from the North American and Eurasian Arctic in 1994 - 98 and from the west coast of Sweden in 1998 contained non-racemic HEPX with mean EFs = 0.663-0.696 *(81)*. This is most likely the result of enantioselective metabolism of heptachlor in soils and subsequent emission of HEPX *(74)*. TC (EF = 0.451 - 0.478) and sometimes CC (EF = 0.503 - 0.513) were also non-racemic in these air samples (Figure 5). On the other hand, TC and CC in surface water sampled under the ice cap in the Canadian Arctic in 1994 were racemic *(20)*, possibly due to residues that date from earlier decades when chlordane was in common use.

Air-Water Exchange

The gas exchange of HCH isomers in the Great Lakes undergoes seasonal changes due to variations in water temperature, air concentrations and stratification of the water column *(13)*. In the western Arctic Ocean and Bering-Chukchi seas, the net exchange direction of α-HCH has reversed in response to declining atmospheric concentrations, from deposition in the 1980s to volatilisation in the 1990s *(19,56,82,83)*. Currently, α-HCH is close to air-water equilibrium in the eastern Arctic Ocean and γ-HCH is still undergoing net deposition *(21)*. One reason for this might be the replacement of technical HCH by lindane, which is currently used in several countries.

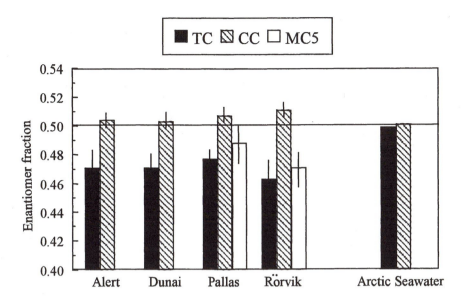

Figure 5. EF values of trans-chlordane (TC), cis-chlordane (CC) and chlordane MC-5 in arctic air and seawater. Vertical lines indicate standard deviations. (Adapted from ref. 81, copyright American Chemical Society).

Volatilization of α-HCH from water bodies can be followed by the appearance of non-racemic α-HCH in the air boundary layer above the water. Air samples were depleted in (+)α-HCH above the unfrozen areas of the western *(19)* and eastern *(21)* Arctic Ocean and in (-)α-HCH above the Bering-Chukchi seas *(19)*, in accord with the enantiomer depletions in the surface waters of these regions. The α-HCH in air over ice-covered portions of the Arctic Ocean were close to racemic, even though non-racemic α-HCH was found in the underlying water. Thus, the racemic signature of α-HCH found in continental regions *(13,16)* is also seen over the Arctic Ocean when gas exchange is hindered.

In Lake Ontario, (+)α-HCH is preferentially depleted (EF = 0.459) *(13)*. The enantiomer profiles in air over the lake showed seasonal changes which followed the trend of water/air fugacity ratios. In fall and spring, fugacity ratios were <1.0 (net deposition) and EFs were close to 0.50 (racemic) indicating that most of the α-HCH came from distant transport. In summer, fugacity ratios were >1.0 (net volatilization) and EFs in air declined to as low as 0.476 as non-racemic α-HCH volatilized from the lake.

Air-water exchange of α- and γ-HCH was investigated in the southern Baltic Sea during summer, 1997 and winter, 1998 *(22)*. The HCHs were close to partitioning equilibrium as judged from their Henry's law constants, but net fluxes varied over time, especially during the summer. Variations in the concentration fraction F_α = α-HCH/(α-HCH + γ-HCH) in the air were related to transport direction, with lower fractions (enrichment of γ-HCH) associated with airflow from central Europe. The EFs of α-HCH in air varied from 0.455 - 0.485 and were generally lower when the air passed over the southern Baltic Sea, where the EF in water averaged 0.445 compared to transport from the north. A linear relationship (r^2 = 0.45, p <0.035) was found between F_α and EF.

An equation for estimating the contribution of a chiral compound from two sources, A and B, in an A-B mixture was derived in terms of the ER values of the compound in A, B, and the mixture *(84)*. The relationship is simplified by using EF instead of ER *(44)*:

$$F_a = (EF_m - EF_b)/ (EF_a - EF_b) \qquad (1)$$

where EF_a, EF_b and EF_m are the EF values for A, B and the mixture. The estimated volatilization contribution of α-HCH to the air over Lake Ontario in a situation where EF_a = 0.459, EF_b = 0.500 and EF_m = 0.476 was 58% *(84)*. The fraction of sea-derived α-HCH in air over the southern Baltic was estimated to be 53-66% in summer and 0-35% in winter *(22)*.

The enantiomeric composition of α-HCH has also been determined in precipitation. Rain samples collected on the shore of Lake Ontario contained racemic α-HCH, even in summer when non-racemic α-HCH was found in air over the lake *(13)*. The authors suggested that rain scavenged racemic α-HCH from above the boundary layer and did not have time to re-equilibrate with the non-racemic α-HCH over the lake. However the EF in rain collected from shore may not be influenced by air over the water but instead dominated by transport of fresh technical HCH *(85)*. Precipitation was collected at the west coast of the Wadden Sea, Germany, in a location where the prevailing winds were from the sea *(85)*. EFs of α-HCH ranged from 0.468 – 0.505 and showed a seasonal trend of lower values (more non-racemic) in the warmer months of the year. Thus, the EF in rain was probably influenced by α-HCH volatilizing from the sea (average EF in water = 0.460), and the apparent contribution of α-HCH from the sea to the rain in summer was estimated to be 29 - 73%.

EFs ranging from 0.478 - 0.494 were found for α-HCH in rain collected from Gotland Island in the southern Baltic *(22)*. In contrast to the German study EFs were more non-racemic in the winter than in summer. However only three rain samples were collected in each season and these were integrated over collection times of 8 - 32 d. The EFs in air samples taken in the region showed a strong influence of transport direction, as noted above.

Hühnerfuss et al. *(54)* demonstrated in laboratory experiments that the dehydrochlorination products of HCHs, the isomeric PCCHs, could also undergo gas exchange. The two isomers appear to have substantially different air/water partition coefficients (Henry's law constants). When air and water were equilibrated in a flask (relative volumes unspecified) for three weeks, 99% of the γ-PCCH partitioned from water into the air, whereas transfer of β-PCCH to air was only minor.

Conclusions and Suggestions for Future Research

The examples presented here illustrate the application of chiral compounds as tracers for following biotic degradation in soil and water, and air-surface exchange of chiral OCPs. Studies involving enantiomers would be enhanced by further work in several areas.

Analytical methods for chiral xenobiotics have been extensively reviewed *(33-35)*. Regarding capillary GC methods, there are presently a limited number of commercial columns which can resolve enantiomers of chiral pesticides. Most chiral columns contain non-bonded phases, an exception being Chirasil-Dex, an immobilized permethylated β-cyclodextrin phase. Non-bonded phases tend to bleed and cause high background signals in GC-ECD and GC-MS. Higher temperature separations, improved signal-to-noise and longer column life could be achieved with bonded-phase chiral columns. A major problem with chiral columns has been variability in stationary phase composition. The cyclodextrins used for chiral stationary phases are often not completely derivatized, and small changes in the percent derivatization can greatly affect chiral separation *(86-89)*. High purity and reproducibility in the preparation of these cyclodextrins are essential for reliable analytical results *(33,35,86-89)*.

Studies of the rates and mechanisms of enantioselective metabolism in soil and water are sparse, and there are very few cases in which attempts have been made to identify/isolate the microbial communities or enzymes responsible. Examples of the latter are the studies in references *(49-52,59)*. An initial study has not shown enantioselectivity of plant enzymes in the degradation of o,p'-DDT *(52)*, however this should not discourage further investigations of this type.

Is it possible that abiotic mechanisms contribute to observed enantioselectivity in soils? Studies have shown selective adsorption of D- or L-amino acids by calcite *(90)*, quartz *(91)* and bentonite clay *(92)* (although the latter is controversial *(93-95)*). There is a need to investigate enantioselective sorption of OCPs by soil constituents such as clay minerals and humics. Indeed,

Lewis et al. *(51)* stated that "...many soils have their genesis with living matter and their organic components may retain chiral centers capable of selectively binding specific enantiomers. Enantioselective sorption to soil organics, if it occurs, could affect the availability of various enantiomers for microbial transformation." Chiral surfactants such as long-chain fatty acids, amino acids and their methyl esters form monolayers at the air-water interface and chiral discrimination is enhanced by bivalent cations (Ca^{+2}, Pb^{+2}, Zn^{+2}) *(96-99)*. Could chiral interactions at the air-water interface lead to discrimination in the gas exchange of enantiomers?

Laboratory or field experiments should be carried out to check whether the enantiomer signatures of chemicals volatilizing from soil and water do in fact match the patterns in these reservoirs, and are not affected by selective binding of one enantiomer. To use enantiomers as tracers of volatilization on a regional scale and incorporate EF data into regional models, it is necessary to determine the spatial and temporal distribution of chiral OCP residues in soil and water.

Acknowledgements

This work was funded in part by the Canadian Toxic Substances Research Initiative (TSRI), a research program managed jointly by Health Canada and Environment Canada, under Project #11. Work in the Canadian Arctic was supported in part by Indian and Northern Affairs Canada under the Northern Contaminants Program. P.A. Helm's work was supported in part by a scholarship from the National Science and Engineering Research Council of Canada.

References

1. Hoff, R.M.; Strachan, W.M.J.; Sweet, C.W.; Chan, C.H.; Shackleton, M.; Bidleman, T.F.; Brice, K.A.; Burniston, D.A.; Cussion, S.; Gatz, D.F.; Harlin, K.; Schroeder, W.H. *Atmos. Environ* **1996**, *30*, 3505-3527.
2. Cortes, D.R.; Basu, I.; Sweet, C.W.; Brice, K.A.; Hoff, R.M.; Hites, R.A. *Environ. Sci. Technol.* **1998**, *32*, 1920-1927.
3. Cortes, D.R.; Hites, R.A. *Environ. Sci. Technol.* **2000**, *34*, 2826-2829.
4. Hillery, B.R.; Simcik, M.F.; Basu, I.; Hoff, R.M.; Strachan, W.M.J.; Burniston, D.; Chan, C.H.; Brice, K.A.; Sweet, C.; Hites, R.A. *Environ. Sci. Technol.* **1998**, *32*, 2216-2221.
5. Simcik, M.F.; Hoff, R.M.; Strachan, W.M.J.; Sweet, C.W.; Basu, I.; Hites, R.A. *Environ. Sci. Technol.* **2000**, *34*, 361-367.

6. Halsall, C. J.; Stern, G. A.; Barrie, L. A.; Fellin, P.; Muir, D. C. G.; Billeck, B. N.; Rovinsky, F. Ya.; Kononov, E. Ya.; Pastukhov, B. *Environ. Pollut.* **1998**, *102*, 51–62.

7. Oehme, M.; Haugen, J.-E.; Schlabach, M. *Environ. Sci. Technol.* **1996**, *30*, 2294-2304.

8. Hung, H.; Halsall, C.J.; Blanchard, P.; Li, H.H.; Fellin, P.; Stern, G.; Rosenberg, B. *Environ. Sci. Technol.* **2002**, *36*, 862-868.

9. Lerche, D.; van der Plassche, E.; Schweigler, A.; Balk, F. *Chemosphere* **2002**, *47*, 617-630.

10. Rodan, B.D.; Pennington, D.W.; Eckley, N.; Boethling, R.S. *Environ. Sci. Technol.* **1999**, *33*, 3482-3488.

11. Williams, A. *Pest. Sci.* **1996**, *46*, 3-9.

12. Hegeman, W.J.M.; Laane, R.W.P.M., *Rev. Environ. Toxicol. Chem.* **2002**, *173*, 85-116.

13. Ridal, J.J.; Bidleman, T.F.; Kerman, B.; Fox, M.E.; Strachan, W.M.J. *Environ. Sci. Technol.* **1997**, *31*, 1940-1945.

14. Harner, T.; Wideman, J.L.; Jantunen, L.M.M.; Bidleman, T.F.; Parkhurst, W.J. *Environ. Pollut.* **1999**, *106*, 323-332.

15. Aigner, E. J.; Leone, A. D.; Falconer, R. L. *Environ. Sci. Technol.* **1998**, *32*, 1162–1168.

16. Leone, A.D.; Amato, S.; Falconer, R.L. *Environ. Sci. Technol.* **2001**, *35*, 4592-4596.

17. Falconer, R.L.; Bidleman, T.F.; Szeto, S.Y. *J. Agric. Food Chem.* **1997**, *45*, 1946-1951.

18. Finizio, A.; Bidleman, T.F.; Szeto, S.Y. *Chemosphere* **1998**, *36*, 345-355.

19. Jantunen, L.M.; Bidleman, T.F. *J. Geophys. Res.* **1996**, *101*, 28837-28846; corrections *Ibid.* **1997**, *102*, 19279-19282.

20. Jantunen, L.M.; Bidleman, T.F. *Arch. Environ. Contam. Toxicol.* **1998**, *35*, 218-228.

21. Harner, T.; Kylin, H.; Bidleman, T.F.; Strachan, W.M.J. *Environ. Sci. Technol.* **1999**, *33*, 1157-1164.

22. Wiberg, K.; Brorström-Lundén, E.; Wängberg, I.; Bidleman, T.F.; Haglund, P. *Environ. Sci. Technol.* **2001**, *35*, 4739-4746.

23. Wiberg, K.; Harner, T.; Wideman, J.L.; Bidleman, T.F. *Chemosphere* **2001**, *45*, 843-848..

24. Wiberg, K.; Letcher, R.J.; Sandau, C.D.; Norstrom, R.J.; Tysklind, M.; Bidleman, T.F. *Environ. Sci. Technol.* **2000**, *34*, 2668-2674.

25. Wiberg, K.; Oehme, M.; Haglund, P.; Karlsson, H.; Olsson, M. *Mar. Pollut. Bull.* **1998**, *36*, 345-353.

26. Fisk, A.T.; Holst, M.; Hobson, K.; Duffe, J.; Moisey, J.; Norstrom, R.J. *Arch. Environ. Contam. Toxicol.* **2002**, *42*, 118-126.

222

27. Fisk, A.T.; Moisey, J.; Hobson, K.A.; Karnovsky, N.J.; Norstrom, R.J. *Environ. Pollut.* **2001**, *113*, 225-238.
28. Moisey, J.; Fisk, A.T.; Hobson, K.A.; Norstrom, R.J. *Environ. Sci. Technol.* **2001**, *35*, 1920-1927.
29. Karlsson, H.; Oehme, M.; Skopp, S.; Burkow, I.C. *Environ. Sci. Technol.* **2000**, *34*, 2126-2130.
30. Tanabe, S.; Kumaran, P.; Iwata, H.; Tatsukawa, R.; Miyazaki, N. *Mar. Pollut. Bull.* **1996**, *32*, 27-31.
31. Iwata, H.; Tanabe, S.; Iida, T.; Baba, N.; Ludwig, J.P.; Tatsukawa, R. *Environ. Sci. Technol.* **1998**, *32*, 2244-2249.
32. Ulrich, E.; Willett, K.L.; Caperell-Grant, A.; Bigsby, R.M.; Hites, R.A. *Environ. Sci. Technol.* **2001**, *35*, 1604-1609.
33. Vetter, W. *Food Rev. Internat.* **2001**, *17*, 113-182.
34. Vetter, W.; Schurig, V. *J. Chromatogr.* **1997**, *774*, 143-175.
35. Kallenborn, R.; Hühnerfuss, H. *Chiral Environmental Pollutants, Trace Analysis and Ecotoxicology*, Springer-Verlag, Berlin, 2001, 209 pages.
36. Mattina, M.J.I.; Iannuchi-Berger, W.; Dykas, L.; Pardus, J. *Environ. Sci. Technol.* **1999**, *33*, 2425-2431.
37. Mattina, M.J.I.; Iannuchi-Berger, W.; Dykas, L. *J. Agric. Food Chem.* **2000**, *48*, 1909-1915.
38. Eitzer, B.D.; Mattina, M.J.I.; Iannuchi-Berger, W. *Environ. Toxicol. Chem.* **2001**, *20*, 2198-2204.
39. Mattina, M.J.I.; White, J.; Eitzer, B.; Iannucci-Berger, W. *Environ. Toxicol. Chem.* **2002**, *21*, 281-288.
40. Szeto, S.Y.; Price, P.M. *J. Agric. Food Chem.* **1991**, 39, 1679-1684.
41. Meijer, S.; Halsall, C.J.; Harner, T.; Peters, A.J.; Ockenden, W.A.; Johnston, A.E.; Jones, K.C. *Environ. Sci. Technol.* **2001**, *35*, 1989-1995.
42. Jantunen, L.M.M.; Bidleman, T.F.; Harner, T.; Parkhurst, W.J. *Environ. Sci. Technol.* **2000**, *34*, 5097-5105.
43. World Health Organisation, *Environmental Health Criteria #38, Heptachlor*, W.H.O., Geneva, 81 pages.
44. Harner, T.; Wiberg, K.; Norstrom, R. *Environ. Sci. Technol.* **2000**, *34*, 218-220.
45. DeGeus, H.-J.; Wester, P.G.; deBoer, J.; Brinkman, U.A.Th. *Chemosphere* **2000**, *41*, 725-727.
46. Müller, M.D.; Schlabach, M.; Oehme, M. *Environ. Sci. Technol.* **1992**, *26*, 566-569.
47. Leone, A.D.; Jantunen, L.M.M.; Harner, T.; Bidleman, T.F. Presented at the Society for Environmental Toxicology and Chemistry (SETAC) National Meeting 2000, Nashville, TN.
48. Vetter, W.; Maruya, K.A. *Environ. Sci. Technol.* **2000**, *34*, 1627-1635.

49. Garrison, A.W.; Schmitt, P.; Martens, D.; Kettrup, A. *Environ. Sci. Technol.* **1996**, *30*, 2449-2455.

50. Buser, H-R., Müller, M.D., Poiger, T., Balmer, M.E. *Environ. Sci. Technol.* **2002**, *36*, 221-226.

51. Lewis, D.L.; Garrison, A.W.; Wommack, K.E.; Wittemore, A.; Steudler, P.; Melillo, J. *Nature* **1999**, *401*, 898-901.

52. Garrison, A.W.; Nzengung, V.A.; Avants, J.K.; Ellington, J.J.; Jones, W.J.; Rennels, D.; Wolfe, N.L. *Environ. Sci. Technol.* **2000**, *34*, 1663-1670..

53. Faller, J.; Hühnerfuss, H.; König, W.; Ludwig, P. *Mar. Pollut. Bull.* **1991**, *22*, 82-86.

54. Hühnerfuss, H.; Faller, J.; König, W.; Ludwig, P. *Environ. Sci. Technol.* **1992**, *26*, 2127-2133.

55. Pfaffenberger, B.; Hühnerfuss, H.; Kallenborn, R.; Köhler-Günther, A.; König, W.; Krüner, G. *Chemosphere* **1992**, *25*, 719-725.

56. Falconer, R.L.; Bidleman, T.F.; Gregor, D.J. *Sci. Total Environ.* **1995**, *160/161*: 65-74.

57. Harner, T.; Jantunen, L.M.M.; Bidleman, T.F.; Kylin, H.; Macdonald, R.W.; Barrie, L.A. *Geophys. Res. Lett.* **2000**, *27*, 1155-1158.

58. Jantunen, L.M.M.; Kylin, H.; Bidleman, T.F. *Deep Sea Res.* **2002**, in press.

59. Ludwig, P.; Hühnerfuss, H.; König, W.; Gunkel, W. *Marine. Chem.* **1992**, *38*, 13-23.

60. Buser, H.-R., Müller, M.D. *Environ. Sci.Technol.* **1995**, *29*, 664-672.

61. Falconer, R.L.; Bidleman, T.F.; Gregor, D.J.; Semkin, R.; Teixeira, C. *Environ. Sci.Technol.* **1995**, *29*, 1297-1302.

62. Helm, P.A.; Diamond, M.; Semkin, R.; Bidleman, T.F. *Environ. Sci.Technol.* **2000**, *34*, 812-818.

63. Alaee, M.; Moore, L.; Wilkinson, R.J.; Spencer, C.; Stephens, G. *Organohalogen Cpds.* **1997**, *31*, 282-285.

64. Law, S.A.; Diamond, M.L.; Helm, P.A.; Jantunen, L.M.M.; Alaee, M. . *Environ. Toxicol. Chem.*. **2001**, *20*, 2690-2698.

65. Law, S.A., Bidleman, T.F., Martin, T., Ruby, M. Paper 236, presented at the Society for Environmental Toxicology and Chemistry national meeting, 2001, Baltimore, MD.

66. Müller, M.D.; Buser, H.-R. *Environ. Sci.Technol.* **1997**, *31*, 1953-1959.

67. Buser, H.-R.; Müller, M.D. *Environ. Sci.Technol.* **1997**, *31*, 1960-1967.

68. Buser, H.-R.; Müller, M.D. *Environ. Sci.Technol.* **1998**, *32*, 626-633.

69. Zipper, C.; Suter, M.J.-F.; Haderlein, S.B., Gruhl, M.; Kohler, H.-P.E. *Environ. Sci.Technol.* **1998**, *32*, 2070-2076.

70. Bucheli, T.D.; Müller, S.R.; Heberle, S.; Schwarzenbach, R.P. *Environ. Sci. Technol.* **1998**, *32*, 3457-3464.

224

71. Bucheli, T.D.; Müller, S.R.; Voegelin, A.; Schwarzenbach, R.P. *Environ. Sci. Technol.* **1998**, *32*, 3465-3471.
72. Buser, H.-R.; Poiger, T.; Müller, M.D. *Environ. Sci.Technol.* **2000**, *34*, 2690-2696.
73. Bethan, B.; Bester, K.; Hühnerfuss, H.; Rimkus, G. *Chemosphere* **1997**, *11*, 2271-2280.
74. Bidleman, T. F.; Jantunen, L. M.; Harner, T.; Wiberg, K.; Wideman, J. L.; Brice, K.; Su, K.; Falconer, R.L.; Aigner, E. J.; Leone, A. D.; Ridal, J .J.; Kerman, B.; Finizio, A.; Alegria, H.; Parkhurst, W. J.; Szeto, S. Y. *Environ. Pollut.* **1998**, *102*, 43–49.
75. Bidleman, T.F.; Falconer, R.L. *Environ. Sci. Technol.* **1999**, *33*, 206A-209A.
76. Bidleman, T.F.; Jantunen, L.M.M.; Wiberg, K.; Harner, T.; Brice, K.A.; Su, K.; Falconer, R.L.; Leone, A.D.; Aigner, E.J.; Parkhurst, W.J. *Environ. Sci. Technol.* **1998**, *32*, 1546-1548.
77. Ulrich, E.M.; Hites, R.A. *Environ. Sci.Technol.* **1998**, *32*, 1870-1874.
78. U.S. Dept. of Health and Human Services *Toxicology Profile for Chlordane* **1992**, Public Health Service, Agency for Toxic Substances and Disease Control, Atlanta, GA, 159 pages + appendices.
79. Leone, A.D.; Ulrich, E.M.; Bodner, C.E.; Falconer, R.L.; Hites, R.A. *Atmos. Environ.* **2000**, *34*, 4131-4138.
80. Spencer, W.F.; Singh, G.; Taylor, C.D.; LeMert, R.A.; Cliath, M.M., Farmer, W.J. *J. Environ. Qual.* **1996**, *25*, 815-821.
81. Bidleman, T.F.; Jantunen, L.M.M.; Helm, P.A.; Brorström-Lundén, E.; Juntto, S. *Environ. Sci.Technol.* **2002**, *36*, 539-544.
82. Jantunen, L.M.; Bidleman, T.F. *Environ. Sci.Technol.* **1995**, *29*, 1081-1089.
83. Bidleman, T.F.; Jantunen, L.M.; Falconer, R.L.; Barrie, L.A. *Geophys. Res. Lett.* **1995**, *22*, 219-222.
84. Bidleman, T.F.; Falconer, R.L. *Environ. Sci.Technol.* **1999**, *33*, 2299-230.
85. Bethen, B.; Dannecker, W.; Gerwig, H.; Hühnerfuss, H.; Schulz, M. *Chemosphere* **2001**, *44*, 591-597.
86. Jaus, A.; Oehme, M. *Chromatographia* **1999**, *50*, 299-304.
87. Jaus, A.; Oehme, M. *Organohalogen Cpds.* **1999**, *40*, 387-390.
88. Vetter, W.; Klobes, U.; Luckas, B.; Hottinger, G. *Chromatographia* **1997**, *45*, 255-262
89. Vetter, W.; Klobes, U.; Luckas, B.; Hottinger, G. *Organohalogen Cpds.* **1998**, *35*, 305-308.
90. Hazen, R.M.; Filley, T.R.; Goodfriend, G.A. *Proc. Nat. Acad. Sci.* **2001**, *98*, 5487-5490.

91. Bonner, W.A.; Kavasmaneck, P.R.; Martin, F.S. *Science* **1974**, *186,* 143-144.
92. Bondy, S.C.; Harrington, M.E. *Science* **1979**, *203,* 1243-1244.
93. Youatt, J.B.; Brown, R.D. *Science* **1981**, *212,* 1145-1146.
94. Wellner, D. *Science* **1979**, *206,* 484
95. Bondy, S.C. *Science* **1979**, *206,* 484.
96. Gericke, A.; Hühnerfuss, H. *Langmuir* **1994**, *10,* 3782-3786.
97. Neumann, V.; Gericke, A.; Hühnerfuss, H. *Langmuir* **1995**, *11,* 2206-2212.
98. Hühnerfuss, H.; Gericke, A.; Neumann, V.; Stine, K.J. *Thin Solid Films* **1996**, *284-285,* 694-697.
99. Hühnerfuss, H.; Neumann, V.; Stine, K.J. *Langmuir* **1996**, *12,* 2561-2569.

Chapter 13

Distributional Risk Assessment for Agrochemicals: Triazine Herbicides

Keith R. Solomon

Department of Environmental Biology and Centre for Toxicology, University of Guelph, Guelph, Ontario N1G 2W1, Canada (ksolomon@tox.uoguelph.ca)

Concern for the environment has resulted in greater scrutiny of both old and new plant protection products and increased efforts have been directed to developing more rigorous but more realistic procedures for the ecotoxicological risk characterization of these agrochemicals. These techniques include a better understanding of the ecological role of organisms and probabilistic analyses of toxicity and exposure data.

The ecological basis for these risk assessments has been broadened though a better understanding of the relationship between structure and function in populations of wildlife and the role of keystone species in maintaining ecosystem functioning. This understanding has been incorporated into risk assessment through a better understanding of population recovery rates and functional resiliency of ecosystems. In assessing the possible effects of atrazine, a triazine herbicide, a number of factors were considered. These include specificity towards plants rather animals, the reversibility of mechanism of action, and the ability of many plant species to recover rapidly.

The use of probabilistic approaches has improved our ability to combine toxicity data for many species into ecological risk assessments. Large toxicity data sets are available for several pesticides such as atrazine.

Distributional analysis of exposure data from US Geological Survey analyses of surface waters and other sources in a number of watersheds allowed estimations of exceedence probabilities and generally showed low risks from these triazine herbicides, even in the more sensitive groups of organisms (plants).

Introduction

Pesticides, like all other chemicals, may have certain combinations of characteristics that lead to the possibility of adverse effects in the environment. These include the intrinsic toxicity of the substance and the potential for exposure of organisms to the substance. The toxicological properties of a substance are determined by its molecular structure and the biochemistry and physiology of the receptor organisms exposed to it. Exposure to the substance is also dependent on interactions between its chemical and physical properties and the environment in which it occurs. In this, pesticides do not possess special properties that make them chemically or physically unique as a class or give them special toxicological or environmental properties that are not also found in other substances. In fact one of the only differences between pesticides and other substances is some degree of specificity of action for the pest organism.

Pesticides are specifically used for the control of unwanted organisms for the protection of crops, human health, or structures. Some form of risk assessment is normally carried out in the process of deciding to use the pesticide, for example, the cost of the pesticide may be a considered in relation to the benefits of its use. Thus a farmer may consider the cost of the pesticide used to control an infestation of insects in a fruit crop against the benefit resulting from increased value of the crop to the ultimate consumer, be this in improved quality, higher yield, or greater value. These types of assessments are familiar to the public and have been used for many years in the process of agricultural production. As the risks (loss of the crop, etc.) are measured in financial units and the cost of the control measures are measured in the same units and simple arithmetic can be used to determine if the risk of use (cost) is worth the benefit (profit).

Although risk assessment decisions are taken as part of the use of pesticides in pest management, these decisions are internalized to the farm, the field, or the structure. These decisions do not necessarily consider effects of the pesticide on non-target organisms outside the area of use or on non-target organisms that utilize the agroecosystem as habitat (birds and mammals). These external risks are those that are focused on in environmental regulations and that raise concerns in the public. These are the subject of this paper and several accompanying papers in this volume.

Basic Ecological Principles

Structure and Function

In contrast to human health protection, individual organisms in the ecosystem are generally regarded as transitory and, because they are usually part of a food chain, are individually expendable (1). Thus, most assessment of ecological risk assessment are directed to the population of organisms, rather than the individual. Only in the case of the rare or endangered species are organisms in the environment provided similar protection to that enjoyed by humans. Generally, ecological risk assessment is aimed at protection of the functions of populations, communities and ecosystems. Even effects on a population may not necessarily be of concern (to the ecosystem) so long as the functions the affected population can be taken over by other organisms in the community. In this context, function is the interaction of the population with other populations or with the surrounding abiotic environment. Functions in ecosystems are normally related to energy and nutrient flow: production of biomass (primary production), consumption of biomass (grazing or predation), controlling the abundance of other (prey) species, providing food to predators, or processing organic detritus, such as shredding plant tissue, macerating animal remains, and mineralizing organic compounds (1).

Redundancy of function is fundamental to the continuance of ecosystems in the face of natural environmental stressors and has likely been selected for by fluctuating and unpredictable environmental conditions. Most ecosystems possess functional redundancy, where multiple species are able to perform indispensable functions (2-4). Functional redundancy is particularly relevant to the ecotoxicological risk assessment. It is the basis for being able to tolerate effects in some more sensitive populations as these are unlikely to impair the functions of the ecosystem as a whole. This is the basis for being able to tolerate some species being affected, such as in setting water quality guidelines (5). There is a general relationship between exposure concentration and impact of any substance, however, there are deviations from this general rule (Figure 1). In this illustration, ecosystem functions may be maintained where few species are affected but, as the number of species affected increases, indirect effects become dominant and the effects of the substance are greater than predicted. Redundancy of function has been observed in a number of experimentally manipulated systems ranging from

230

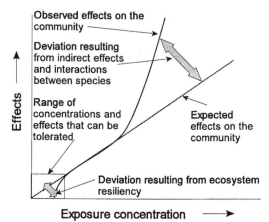

Figure 1 Illustration of ecosystem resiliency
in response to pesticides and effects caused
by interactions between organisms (adapted
from a drawing by D. Hansen.

terrestrial (6, 7) to aquatic (8-13). These observations support the concept that, in ecotoxicological risk assessment, some effects at the level of the organism and population can be allowed, provided that these effects are restricted on the spatial and temporal scale. In other words, they do not affect all communities all of the time and that keystone organisms are not adversely affected. In the context of selecting assessment measures, it has become increasingly recognized that these should be at the functional level of populations and the community and that some effects on populations and species diversity may thus be tolerated.

Effects in keystone species are less easily tolerated. Keystone species often supply physical habitat or modify the habitat in a way that cannot be replicated. Thus, removal of habitat (populations of other species such as plants) may be the root cause of the risk to a population designated for protection. An example of this is seen in the case of the spotted owl in the Pacific NW and removal of its nesting habitat (14). Similarly, effects of pulp mill effluents on larval and juvenile fish in the Baltic is another example of the importance of habitat (15) as is the relationship between sea urchins, kelp, and sea otters (16). Examples of these types of keystone responses are infrequently reported for pesticides in the non-agricultural environment but are a very real problem in integrated pest management (IPM) situations within the agroecosystem.

Frequency of Effects

How often effects occur is a very important consideration in assessing ecological risks relation to the ecological cost of recovery from the event (17). When assessing exposures, the return frequency protected against should be compatible with the resiliency of sensitive populations judged to be important and in need of protection. Resiliency is determined by life cycle characteristics and the reproductive capacity of the potentially affected organisms and the ability of their populations (or their function in the ecosystem) to recover from the toxic event. This issue should be addressed early in the problem formulation stage of the risk assessment as it affects the exposure and the effects analysis. The Aquatic Risk Assessment and Mitigation Dialogue Group (18) recommended conservative approaches to probabilistic ecological risk assessment (PERA), such as the use of low return frequencies (i.e., one or fewer occurrences in thirty years). This safeguards all organisms in situations where limited information is available on the mode of action or the sensitivity of species. Where better information is available, more appropriate return frequencies may be used. For example, more frequent exposure to a stressor may be tolerated where a stressor affects organisms with short life cycles and high rates of reproduction, such as algae, some zooplankton, etc. In temperate regions, many ecosystems undergo a period of dormancy and the system is, in a sense, reset seasonally by the winter. Thus, for some organisms, mechanisms for propagation beyond the winter reset already exist, and dormant stages are produced from which populations will develop in the next season. Similar mechanisms exist in environments with a dry season where ephemeral water bodies are subjected to drying out. Therefore, as many organisms in these regions undergo seasonal resets, a stressor return that occurs less frequently than once per season is likely to be tolerable from the viewpoint of the long-term productivity of the population and the sustainability of function in the ecosystem, especially if the effects are spatially restricted and repopulation can occur from refugia. However, protection of longer-lived species without seasonal resets, such as some fish, birds, or mammals, may require the consideration of return frequencies of several years. From a practical point of view, the annual maximum concentration of the stressor is a useful starting point for the purposes of PERA, as it is a conservative return frequency for most organisms. The likelihood of annual maxima exceeding a toxicity threshold may then be assessed using a fixed centile from the species sensitivity distribution or an exceedence profile. As has been pointed out (19), shorter return frequencies can be tolerated if organisms with different reproductive strategies are the most likely to be affected. For example, unicellular algae are among the most sensitive aquatic organisms to atrazine. These algae have rapid reproduction times and can quickly recover from effects, even if these were to occur with return frequencies of 30 days (8). Similarly, many zooplankton species can recover from reductions in numbers of several orders of magnitude within a month or two (9, 20). From a practical point of view, monthly return frequencies may be more important in assessing risks in these types of organisms.

Along with the consideration of return frequency, the time between events may also be important. This issue was specifically addressed in the development of the RADAR post-processing tool (19). Again, it is well known that organisms may recover from exposure to a toxic substance if this substance does not cause mortality but merely acts as a stasis agent. For example, atrazine and the other triazine herbicides are reversible inhibitors of photosynthesis, and when an exposed plant is moved to uncontaminated environment, it will continue to grow and develop (21, 22), provided that the exposure time to the herbicide has not been so long that its internal energy reserves have been depleted to the point of lethality. Similar responses have been observed in chronically exposed fish when they are moved to uncontaminated media (23). Standardized toxicity assays for algae are based on growth as an endpoint, and concentrations of a substance that inhibit growth are often assumed to cause death. Concentrations of stressors that result in death have been shown to be different from those that result in the reduction of growth (24). Practically, for stressors such as these, PERA should be focused on the likelihood of concentrations exceeding a certain threshold value for a certain time and then followed by a time interval judged to be inadequate to allow for recovery to occur.

In practice, the likelihood of occurrence of exposures that meet certain criteria can be assessed. These criteria may, for example, consist of an effect concentration threshold, a duration, and an interval between exposures. As illustrated for a measured data set for atrazine in Lost Creek, Ohio (17), no exposures exceeded the threshold of 5.6 g/L for a period of more than 6 days, and many of these were followed by a considerable period of exposure less than the threshold. Comparison of these exposures to laboratory studies with algae that demonstrated full recovery after up to 14 days of exposure to atrazine that caused stasis of cell growth (22) suggested insignificant risk to algal populations in this particular situation; however, the example illustrates how the method could be used.

Probabilistic Risk Assessment

Paracelsus (1493-1541) was the first to state the observation that "The dose makes the poison" an utterance that has become one of the central dogmas of toxicology[1]. Thus the assessment of poisoning or hazard requires a knowledge of both the exposure and the inherent toxicity of the chemical. This principle has formed the basis for the concepts of risk assessment for pesticides such as have been discussed in a number of reports (18, 25-28) and books (1, 29). The EPA Framework for Risk Assessment (25, 26) serves as a useful general model.

Probabilistic approaches to risk assessment have been suggested for upper tiers in the ecological risk assessment process (18, 19). These methods use distributions of species sensitivity combined with distributions of exposure concentrations to better describe the likelihood of exceedences of effect thresholds and thus the risk of adverse effects.

The probability of occurrence of a particular event is, and has been, widely used in the characterization of risk from many physical and medical events in humans (the insurance industry) and for protection against failure in mechanical and civil engineering projects (time between failures, one-in-one-hundred-year floods, etc.).

Using distributions of species sensitivities is not a new idea. Distributional approaches have been used in the regulation of food additives (30) for the protection of human health for several years. From the environmental point of view, distributions of LC50s were used to distinguish between resistant and susceptible populations of animal ectoparasites (31) and setting environmental quality guidelines (5, 32). Other authors who have expanded upon the probabilistic risk estimation process include (33-38). The suggestion to compare distributions of species sensitivity directly to distributions of exposure concentrations (39) was recommended for pesticide risk assessment by the Aquatic Risk Assessment Dialogue Group (18), demonstrated for metals and other substances (40), and incorporated in a computer program (41). Probabilistic ecological risk assessment has been used in a number of risk assessments for pesticides and other substances (8, 13, 42-44), and has been recommended for regulatory risk assessment of pesticides (19). The general concepts as they apply to ecological and human health risk assessment have been reviewed and extensively discussed (1, 17, 45-53).

The principle of probabilistic approach has been described (17, 18, 39, 40, 42) and is illustrated diagrammatically in Figure 2. From cumulative frequency distributions from properly sampled sub-sample of events, it is

---

[1] He actually stated that "All substances are poisons: there is none which is not a poison. The right dose differentiates a poison and a remedy".

possible to assign probabilities to the likelihood that a measure will exceed a certain value. This principle can be applied concentrations of substances in the environment with due consideration for the fact that these data are usually censored by the limits of analytical detection (Figure 2-A). When plotted as a cumulative frequency distribution using a probability scale on the Y axis as a function of $\log_{10}$ concentration (Figure 2-B), these distributions approximate a straight line which can be used to estimate the likelihood that a particular concentration of the substance will be exceeded in an environment where the circumstances controlling the fate and concentration of the substance are similar to the sampled environment. A similar approach can be taken with susceptibility of organisms to the substance (Figure 2-C and D). The combination of these distributions in the probabilistic characterization of risk is illustrated in Figure 2-E. In this procedure, it is assumed that the distri butions of sensitivity represent the range of responses that are likely to be encountered in the ecosystems where the exposures occur (18, 19). If the exposure data were collected over time at a particular site, the degree of overlap of the exposure distribution with the effects distribution can be used to estimate the joint probability of exposure and toxicity, leading to estimates of exceedence probabilities for responses at a fixed effect assessment criterion, such as, for example, the concentration equivalent to the 10th centile of all species or a distribution of inherently more sensitive species (8). Another method of presenting these joint probabilities is in the form of a joint probability curve or exceedence profile (EP). This format was suggested in the AERA program (41), recommended for displaying risks by the Aquatic Working Group of ECOFRAM (19), and has been used to display risks for pesticides such as chlorpyrifos (13) and diazinon (11). The derivation of the EP (Figure 3) is relatively simple and offers a useful tool for site-specific communication of risks as it allows what-if questions to be addressed and gives the risk assessor and risk manager a method for assessing the effects of changes in assumptions, such as the choice of a different centile from the species (13, 19). This can be applied to a number of exposure data sets and the resulting probabilities used for priority setting or in further assessing ecological relevance.

The Triazine Herbicides

The triazine herbicides offer a useful example. Atrazine relatively water soluble and is widely detected in surface waters. It is not bioaccumulative or biomagnified, the mechanism of action well understood and it is considerably more toxic to plants than animals. The mechanism of action is reversible in plants and animals and there is a good database on toxicity

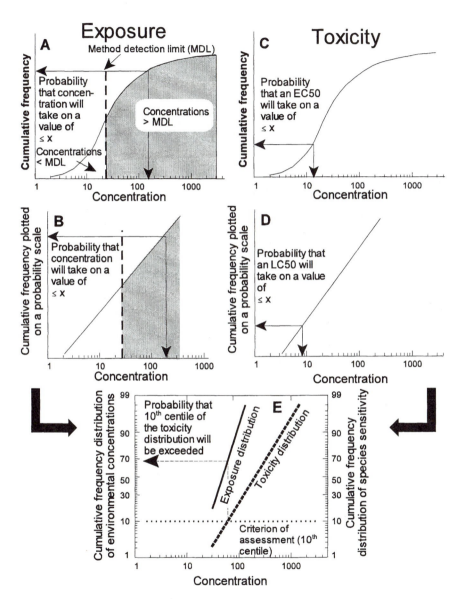

Figure 2 Illustration of the principle of the probabilistic approach (adapted from 54).

236

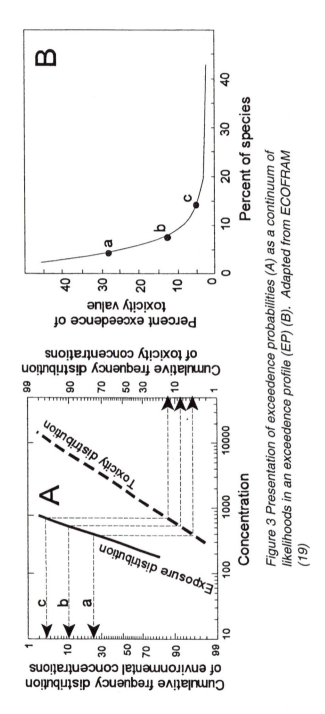

Figure 3 Presentation of exceedence probabilities (A) as a continuum of likelihoods in an exceedence profile (EP) (B). Adapted from ECOFRAM (19)

and data on concentrations in surface waters. Thus, from a basic understanding of the mechanism of action of a pesticide, it was possible to identify and group sensitive organisms, in this case plants (8). This was helpful from the point of view of risk assessment as it allowed the assessment to focus on the groups at higher risk. In addition, with a knowledge of the ecology of the potentially impacted system and observations from microcosm and mesocosm studies with atrazine, it was possible to assess the likelihood of indirect effects as a result of an response in a keystone groups of predator or prey/food organisms (8).

In this example, unicellular algae were amongst the most sensitive aquatic organisms to atrazine. These algae have rapid reproduction times and can quickly recover from reversible effects, even if these were to occur with return frequencies of the order of 30 d (8). From a practical point of view, monthly return frequencies may be more important in assessing risks in these types of organisms.

Along with the consideration of return frequency, the interinterval time between events may also be important. This issue was specifically addressed in the development of the RADAR post-processing tool by ECOFRAM (19). Again, it is well known that organisms may recover from exposure to a toxic substance if this substance does not cause mortality but merely acts as a stasis agent such as is the case with atrazine. Standardized toxicity assays for algae are based on growth as an endpoint and this is often interpreted as resulting from death, however, this may not be the case. Concentrations of stressors that result in death have been shown to be different from those that result in reduction of growth (24). Practically, for stressors such as atrazine, probabilistic risk assessment should be focused on the likelihood of concentration exceeding a certain concentration for a certain time and then followed by a time interval judged to be inadequate to allow for recovery to occur.

In practice, the likelihood of occurrence of exposures that meet certain criteria can be assessed. These criteria may, for example, consist of an effect concentration threshold, a duration, and an interval between exposures. As illustrated for a measured data set for atrazine in Lost Creek, OH (17), no atrazine concentrations exceeded a probabilistically derived chronic 10th centile concentration of 5.6 µg/L (8) for a period of more than 6 days and many of these were followed by a considerable period of exposure concentrations smaller than the threshold. Comparison of these exposures to laboratory studies with algae showing that they could recover fully after up to 14 d of exposure to atrazine (22), suggested insignificant risk to algal populations in this particular situation.

238

References

1. Suter, G.W., Barnthouse, L.W., Bartell, S.M., Mill, T., Mackay, D., and Patterson, S. Ecological Risk Assessment, **1993**, Lewis Publishers,, Boca Raton, FL.
2. Baskin, Y. *Bioscience.* **1994**, *44*, 657-660.
3. Walker, B. *Conserv. Biol.* **1992**, *6*, 18-23.
4. Walker, B. *Conserv. Biol.* **1995**, *9*, 747-752.
5. Stephan, C.E., Mount, D.I., Hansen, D.J., Gentile, J.H., Chapman, G.A., and Brungs, W.A., **1985**, pp. 1-97, US EPA ORD ERL, Duluth MN.
6. Tilman, D. *Ecology.* **1996**, *77*, 350-363.
7. Tilman, D., Wedlin, D., and Knops, J. *Nature.* **1996**, *379*, 718-720.
8. Solomon, K.R., Baker, D.B., Richards, P., Dixon, K.R., Klaine, S.J., La Point, T.W., Kendall, R.J., Giddings, J.M., Giesy, J.P., Hall, L.W.J., Weisskopf, C., and Williams, M. *Environ. Toxicol. Chem.* **1996**, *15*, 31-76.
9. Stephenson, G.L., Kaushik, N.K., Solomon, K.R., Day, K.E., and Hamilton, P. *Environ. Toxicol. Chem.* **1986**, *5*, 587-603.
10. Giddings, J.M., Biever, R.C., Annunziato, M.F., and Hosmer, A.J. *Environ. Toxicol. Chem.* **1996**, *15*, 618-629.
11. Giddings, J.M., Hall, L.W.J., and Solomon, K.R. *Risk Anal.* **2000**, *20*, 545-572.
12. Giddings, J.M., Solomon, K.R., and Maund, S.J. *Environ. Toxicol. Chem.* **2001**, *20*, 660-668.
13. Giesy, J.P., Solomon, K.R., Coates, J.R., Dixon, K.R., Giddings, J.M., and Kenaga, E.E. *Rev. Environ. Contam. Toxicol.* **1999**, *160*, 1-129.
14. James, F.C. *Bull. Ecol. Soc. Am.* **1994**, *June, 1994*, 69-75.
15. Lehtinen, K.-J., Axelsson, B., Kringstad, K., and Strömberg, L. *Nordic Pulp Pap. Res. J.* **1991**, *2*, 81-88.
16. Estes, J.A., Tinker, M.T., Williams, D., and Doak, D.F. *Science.* **1998**, *282*, 473-476.
17. Solomon, K.R. *Risk Anal.* **1996**, *16*, 627-633.
18. SETAC Society for Environmental Toxicology and Chemistry. Pesticide Risk and Mitigation. Final Report of the Aquatic Risk Assessment and Mitigation Dialog Group, **1994**, SETAC Foundation for Environmental Education,, Pensacola, FL, USA.
19. ECOFRAM, **1999**, *Vol.* 1999, USEPA.
20. Kaushik, N.K., Stephenson, G.L., Solomon, K.R., and Day, K.E. *Canadian Journal of Fisheries and Aquatic Science* **1985**, *42*, 77-85.

239

21. Jensen, K.I.N., Stephenson, G.R., and Hunt, L.A. *Weed. Sci.* **1977**, *25*, 212-220.

22. Klaine, S.J., Dixon, K.R., and Florian, J.D., **1996**, pp. 1-92, Novartis Crop Protection, Greensboro, NC.

23. Whale, G.F., Sheahan, D.A., and Kirby, M.F. *in* Sublethal and Chronic Effects of Pollutants on Freshwater Fish **1994**, pp. 175-187, Fishing News (Books) Ltd, London, UK.

24. Faber, M.J., Smith, L.M.J., Boermans, H.J., Stephenson, G.R., Thompson, D.G., and Solomon, K.R. *Environ. Toxicol. Chem.* **1997**, *16*, 1059-1067.

25. USEPA, **1992**, USEPA, Washington, DC, USA.

26. USEPA, **1998**, Risk Assessment Forum, U.S. Environmental Protection Agency, Washington, DC, USA.

27. NRC Issues in Risk Assessment, **1993**, National Academy Press,, Washington, DC.

28. Environment Canada, **1997**, Environment Canada, Ottawa.

29. Reinert, K.H., Bartell, S.M., and Biddinger, G.R. (Eds.) Ecological Risk Assessment Decision-Support System: A Conceptual Design., **1998**, SETAC Press, SETAC Press, Pensacola.

30. Munro, I.C., Rapporteur, Higginson, J., Krewski, D., Pegg, A.E., Rosenkranz, H., Solomon, K.R., Weisburger, E., Williams, G.M., and Wogan, G.N. *Reg. Toxicol. Pharmacol.* **1990**, *12*, 2-12.

31. Solomon, K.R., Baker, M.K., Heyne, H., and van Kleef, J. *Onderstepoort. J. Vet. Res.* **1979**, *46*, 171-177.

32. Kooijman, S.A.L.M. *Water. Res.* **1987**, *21*, 269-276.

33. Van Straalen, N.M., and Denneman, C.A.J. *Ecotox. Environ. Safe.* **1989**, *18*, 241-251.

34. Wagner, C., and Løkke, H. *Water. Res.* **1991**, *25*, 1237-1242.

35. Aldenberg, T., and Slob, W. *Ecotox. Environ. Safe.* **1991**, *18*, 221-251.

36. Aldenberg, T., and Slob, W. *Ecotox. Environ. Safe.* **1993**, *25*, 48-63.

37. Okkerman, P.C., Plassche, E.J., Sloof, W., Van Leeuwen, C.J., and Canton, J.H. *Ecotox. Environ. Safe.* **1991**, *21*, 182-193.

38. Okkerman, P.C., van der Plassche, E.J., Emans, H.J.B., and Canton, J.H. *Ecotox. Environ. Safe.* **1993**, *25*, 341-359.

39. Cardwell, R.D., Parkhurst, B.R., Warren-Hicks, W., and Volosin, J.S. *Water Environ. Technol.* **1993**, *5*, 47-51.

40. Parkhurst, B.R., Warren-Hicks, W.J., Cardwell, R.D., Volosin, J.S., Etchison, T., Butcher, J.B., and Covington, S.M., **1996**, Water Environment Research Foundation, Alexandria, VA, USA.

41. The Cadmus Group Incorporated, **1996**, Water Environment Research Foundation, Alexandria, VA.

42. Klaine, S.J., Cobb, G.P., Dickerson, R.L., Dixon, K.R., Kendall, R.J., Smith, E.E., and Solomon, K.R. *Environ. Toxicol. Chem.* **1996,** *15,* 21-30.

43. Hall, L.W.J., Giddings, J.M., Solomon, K.R., and Balcomb, R. *Crit. Rev. Toxicol.* **1999,** *29,* 367-437.

44. Cardwell, R.D., Brancato, M.S., Toll, J., DeForest, D., and Tear, L. *Environ. Toxicol. Chem.* **1999,** *18,* 567-577.

45. Forbes, T.L., and Forbes, V.E. *Funct. Ecol.* **1993,** *7,* 249-254.

46. Forbes, V.E., and Forbes, T.L. Ecotoxicology in Theory and Practice, **1994,** Chapman and Hall.

47. Balk, F., Okkerman, P.C., and Dogger, J.W., **1995,** Organization for Economic Co-operation and Development (OECD), Paris.

48. Richardson, G.M. *Human Ecol. Risk Assess.* **1996,** *2,* 44-54.

49. Anderson, P.S., and Yuhas, A.L. *Human Ecol. Risk Assess.* **1996,** *2,* 55-58.

50. Burmaster, D.E. *Human Ecol. Risk Assess.* **1996,** *2,* 35-43.

51. Power, M., and McCarty, L.S. *Human Ecol. Risk Assess.* **1996,** *2,* 30-34.

52. Bier, V.M. *Risk Anal.* **1999,** *19,* 703-709.

53. Roberts, S.M. *Human Ecol. Risk Assess.* **1999,** *5,* 729-736.

54. Solomon, K.R., and Chappel, M.J. *in* Triazine Risk Assessment **1998,** (Ballantine, L., McFarland, J., and Hackett, D., Eds.), Vol. 683, pp. 357-368, American Chemical Society, Washington, DC, USA.

Chapter 14

Deposition of Pesticides in Riparian Buffer Zones Following Aerial Applications to Christmas Tree Plantations

Allan S. Felsot[1], Steve L. Foss[2], and Jianbo Yu[3]

[1]Food and Environmental Quality Laboratory, Washington State University, 2710 University Drive, Richland, WA 99352–1671
[2]Pesticide Management Division, Washington State Department of Agriculture, P.O. Box 42589, Olympia, WA 98504–2589
[3]Environmental Health Laboratory, Department of Environmental Health, University of Washington, Seattle, WA 98195–7234

No-spray buffer zones are often required in forestry operations to prevent direct entry of aerially applied pesticides into aquatic resources. The buffer zones often overlap with riparian management zones, and best management practices are recommended to minimize deposition in these areas. Little research has been conducted to characterize deposition of pesticides in riparian zones and to validate required no-spray buffer zone widths. We tested the hypothesis that riparian vegetation surrounding Christmas tree plantations would filter pesticide residues, thereby effectively minimizing contamination of streams or ponds. Deposition of chlorothalonil and endosulfan along two transects in each of two Christmas tree plantations in Lewis County, Washington were monitored following applications by helicopter. Deposition was also simulated using the tier III forestry module of AgDRIFT®. No-spray buffer zones of 15.2 and 91.5 m for chlorothalonil and endosulfan applications, respectively, were delineated around a flowing stream and a pond. Residues of both pesticides were detected in the inner area of riparian management zones at low levels (0.01% or less of the theoretical application rate). Chlorothalonil was detected in one water sample (<1 ppb). AgDRIFT simulated deposition of residues at the water edge reasonably well but sometimes over predicted residues by 10-fold. AgDRIFT simulations were very sensitive to humidity. Even when AgDRIFT over predicted residues, the slope of decline of residues with distance paralleled the slope of the deposition curve for measured residues. Further deposition studies and validation of AgDRIFT will facilitate risk assessments for the effects of pesticide deposition in riparian zones.

Sensitive aquatic and terrestrial habitats can be protected by establishing buffer zones around them to restrict activities that potentially degrade their integrity. Buffers are usually permanently vegetated areas consisting of native and/or non-native species of trees, grasses, and wetland plants (1). Buffers have been most studied as management tools for protection of aquatic resources from runoff, erosion, and subsurface transport of nonpoint agriculturally derived pollutants including nutrients and pesticides (2). Such protection may be required over long periods of time following application of agricultural chemicals.

Buffers may also be effective for protecting sensitive terrestrial or aquatic habitats from pesticides during periods of application, i.e., point sources of contamination, and therefore would be useful during a very short period of time. In agroecosystems, the buffer might be considered as a temporary minimum distance between application to a target crop and a nontarget crop, aquatic resource, or residence. In the case of pesticide applications in forestry, a buffer may be considered as a permanent space of defined ecological structure surrounding a stream or river. Indeed, the buffer can overlap with the riparian management zone but may include an additional setback distance.

Although minimization of off-site contamination would be a key objective for the designation of any buffer protecting aquatic resources, the numerical or narrative criteria for protection of aquatic biota may be an important factor to consider in determining appropriately sized buffer zones for pesticide applications. In this case, the type of pesticide and its mode of application would influence the size of the buffer. For example, under authority of the Clean Water Act, the U.S. EPA has established for several insecticides ecologically protective water quality criteria significantly below 0.1 μg L^{-1} (ppb) (3). The comparatively high toxicity of organophosphorus and chlorinated cyclodiene insecticides to aquatic organisms has resulted in water quality criteria as low as 0.009 ppb for diazinon and 0.056 ppb for endosulfan. On the other hand, herbicides are generally less hazardous than insecticides to aquatic organisms. In Canada, the criteria for atrazine has been set to 2 ppb (4). To meet any water quality criteria, the buffer zone should be designed based on the likely pathway of aquatic entry.

In both agricultural and forestry applications, the buffers for mitigating drift from pesticide applications are essentially no-spray zones. The U.S. Environmental Protection Agency (EPA) has proposed that pesticide product labels impose no-spray zones to protect habitat and people (5). A no-spray zone was defined as "an area in which direct application of the pesticide is prohibited; this area is specified in distance between the closest point of direct pesticide application and the nearest boundary of a site to be protected, unless otherwise specified on a product label." The size of no-spray zones for the most part has been left indeterminate, but some products already require specific buffers. For example, endosulfan has been associated with a comparatively high incidence of fish kills and labels were modified in 1992 with language stating, "Due to the risk of runoff and drift, do no apply within a distance of 300 feet [91 m] of lakes, ponds, streams and estuaries" (6).

Various state regulations may also routinely specify no-spray zones. For example, the State of Washington Administrate Code regulates the use of chemicals in forests and requires best management practices for application of pesticides (7). The rules dictate the size of buffer zones or setbacks from bodies of water. The buffer or no-spray zone overlaps the riparian management zone and is measured along a horizontal distance from the bank-full width of the body of water (called the "edge") through a vegetated area known as the inner zone. Inner zone sizes vary by site classification and generally range from 24-41 m as measured in a horizontal distance from the water's edge (8). Depending on inner zone size, general weather conditions, nozzle types (which control spray coarseness), and spray release heights, the buffer zone sizes for the majority of perennial and seasonal bodies of water range from a low of 24 m under optimal spraying conditions (e.g., winds between 4.8-10 km h^{-1}) to 99 m under sub optimal conditions (e.g., winds calm or blowing in a direction toward a sensitive habitat).

Several research studies have attempted to define buffer widths that would protect aquatic organisms and native vegetation from drift of specific pesticides, including asulam (9), glyphosate (10,11,12), endosulfan (13), cypermethrin (14,15), fenvalerate (16), and permethrin (17,18). Studied habitats included both forestry plantations and agricultural fields.

Specific regulatory-prescribed no-spray zones have been infrequently validated or tested for effectiveness. Research in an Oregon forest showed that a 10-m buffer around streams prevented entry of the herbicide glyphosate that was applied by helicopter (19). In Washington State, however, 15-m buffers around streams adjacent to forestry and Christmas tree plantations did not prevent entry of herbicide residues during helicopter applications (20). Residues did not exceed established water quality criteria and dissipated within 24 hours. Similarly, residues in forest streams in Oregon that were unintentionally oversprayed also dissipated to undetectable levels within 24 hours (19).

In Washington State, best management practices are required to minimize entry of pesticides applied aerially into the inner zones of riparian areas of forest trees, except Christmas trees that are cultivated by agricultural methods (7).

Although studies have attempted to define buffer zones to protect aquatic organisms and native vegetation, few have focused on the extent of deposition within wooded riparian management zones that serve as buffers to streams. To simultaneously address the effectiveness of prescribed buffer widths for forest chemical applications and characterize the deposition of residues within riparian management zones, we monitored pesticide residue deposition at two Christmas tree plantations during commercial spraying operations. We hypothesized that wooded riparian management zones surrounding Christmas tree plantations could serve as effective filters that minimize drift of pesticide residues into water during application by helicopters, the predominant mode of application for larger commercial operations. To test this hypothesis, we monitored spray deposits under the helicopter swaths within the fields, the adjacent buffers surrounding the riparian management zone, and within the riparian zone itself. We also collected water from the buffered streams at different time intervals before and following pesticide applications. The results of deposition monitoring were also compared to predictions from the model AgDRIFT® (21).

Methods

Site Description

Pesticide deposition during commercial spraying operations was monitored at two Christmas tree plantations in western Washington during 1994 (Figures 1, 2). At the time of this study (1994), all bodies of water adjacent to or on the plantations were required by the Washington State Forest Practice Rules to be protected by a no-spray buffer zone during aerial application.

Site 1 was located south of Winlock in Lewis County. The sprayed area consisted of about 21 ha of Douglas fir Christmas trees growing on both sides of Duffy Creek, which was buffered by a riparian management zone (RMZ) (Figure 1). This zone consisted of Douglas fir, western hemlock, and western red cedar (dbh, diameter at breast height, 10-76 cm); hardwoods (red alder, big leaf maple and black cottonwood – dbh 5 to 56 cm); and groundcover (sword fern, devils club and salmonberry). The terrain was hilly throughout the plantation and the RMZ sloped steeply toward the creek.

Site 2, located near Silver Creek in Lewis County consisted of a Noble fir Christmas tree plantation located on the eastern side of an approximately 3-ha pond (Figure 2). The terrain was very hilly and part of the plantation sloped steeply toward the eastern edge of the pond. The shoreline vegetation along the eastern edge of the pond was dominated by non-native reed canary grass (0.6-1.5 m tall). About 20 percent of the buffer was stocked with red alder and black cottonwood (dhh 15-46 cm). The northern end of the pond was bordered primarily by alder trees 15-24 m tall.

Pesticide Application

Several weeks prior to the start of the monitoring studies, the Christmas tree plantation at site 1 was treated with atrazine for weed control. Pesticide monitoring in this study focused on deposition of the insecticide endosulfan and the fungicide chlorothalonil. Christmas trees are susceptible to attack by the Cooley spruce gall adelgid, and infestations at the time of this study (1994) were controlled with endosulfan. Christmas trees are also susceptible to needle blights that were managed with chlorothalonil.

All applications were applied from a Bell Soloy 47 helicopter, flying 2.4 m above the trees at 48 km h^{-1}. The boom length was 9.5 m and held 31 D8 (core 46) nozzles that were oriented 45° back and operated with a pressure of 124 kPa. Wind at both sites at the time of application was calm ($<$1.6 km h^{-1}) and variable in direction. Average temperature was 10°C at both sites. The highest relative humidity (R.H.) was 98% at Site 2 but not measured at Site 1.

During the morning of May 19, 1994 (08:33 – 09:10 h), Site 1 was sprayed on both sides of Duffy creek allowing at least a 15.2-m buffer for application of the fungicide chlorothalonil and 91.5-m buffer for endosulfan (Figure 1). The buffers were actually no-spray zones within the plantation but for endosulfan they extended into the RMZ. Spray swaths near plantation borders were made parallel to the RMZ. Pesticides were mixed in water and applied in six separate loads at a volume of 90.4 L/ha. Loads 1-5 contained chlorothalonil (Agronil 500) applied at a rate of 2.33 kg active ingredient (a.i.)/ha. Loads 2-6 contained endosulfan (Thiodan 3EC) applied at a rate of 0.55 kg a.i./ha. Loads 1 and 2 also contained the drift control adjuvant, Sta-Put, applied at a rate of 0.12 L product/ha. Theoretical surface area deposition rates of chlorothalonil and endosulfan were 23.4 μg cm^{-2} and 5.52 μg cm^{-2}, respectively.

During the morning of June 15, 1994 (07:12 – 07:35 h), about 10 ha of Noble fir Christmas trees at Site 2 were treated in three loads with chlorothalonil (Bravo 720, 3.36 kg a.i./ha, loads 1-3) and endosulfan (Thiodan 3EC, 0.85 kg a.i./ha, loads 2-3) (Figure 2). A non-ionic adjuvant (Kinetic) was added to load 1 and applied at a rate of 0.12 L product /ha. A no-spray buffer extended 15.2 m from the eastern edge of the pond into the trees for the chlorothalonil sprays. For the endosulfan sprays, the buffer was extended 91.5 m into the plantation. The northern side of the plantation was buffered from both pesticide sprays to a distance of approximately 61 m. As at Site 1, initial spray swaths were laid down parallel to the length of the buffer zones. Theoretical surface area deposition rates of chlorothalonil and endosulfan were 33.6 μg cm^{-2} and 8.41 μg cm^{-2}, respectively.

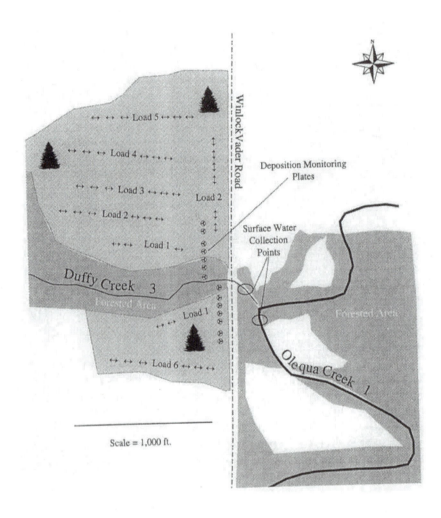

Figure 1. Schematic of Christmas tree plantation at Site 1 near Winlock, WA

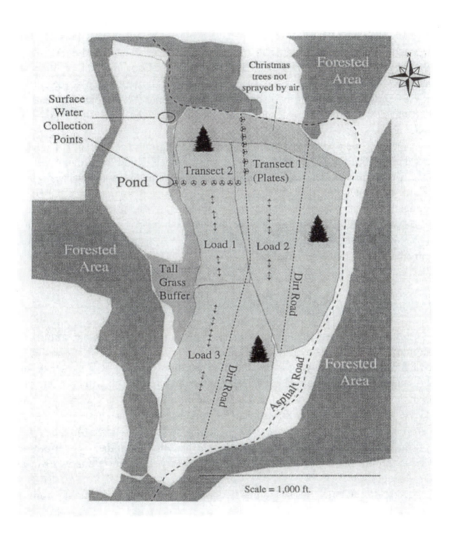

Figure 2. Schematic of Site 2 near Silver Creek, WA

Deposition Monitoring

Flexible aluminum-backed thin layer silica gel plates were used to trap depositing pesticide residues during spraying. Silica gel plates have been used in other atmospheric deposition studies to trap atrazine (22) and chlorpyrifos (23). At each site the plates (10 x 20 cm) were placed at varying distances along transects roughly perpendicular to the long axis of the aquatic body (Figures 1, 2). The transects was started at the bank of the water body and ended within the plantation under the border spray swath. At site 1, a transect was placed on each side of Duffy Creek. The two transects placed at site 2 were perpendicular to each other, with only one oriented toward the pond and the other running parallel to the pond.

At Site 1 the plates were placed directly on the ground at intervals of 3.1-7.6 m along the transect. At site 2, plates along transects 1 and 2 were also placed on 1.5 m high wooden poles. Plates were placed in the open where foliage would not intercept depositing aerosols. Within 4 hours after pesticide application, plates were collected and wrapped individually in aluminum foil. Plates were held on ice during transport to the Washington State University Food & Environmental Quality Lab in Richland, WA where they were stored immediately at -20°C until analyzed.

Analytical Methods

Samples were removed from the freezer, warmed to room temperature, and cut into 2 x 5 cm strips over sheets of aluminum foil. The strips were transferred to a 250 mL Erlenmeyer flask. The scissors used to cut the plates and the aluminum foil which trapped particles of silica gel were rinsed with acetone into the extraction flask. Chlorothalonil and endosulfan were extracted in 75 mL of 1:1 v/v acetone-hexane by shaking for 60 minutes on a rotary shaker. Solvents were decanted through baked (110°C) sodium sulfate in a glass funnel into a 100-mL volumetric flask, and acetone was added to volume. Dilutions or concentrations of the extract were made as necessary to optimize analysis by gas chromatography.

Each day a batch of field-collected plates was extracted, quality control (QC) samples were also analyzed. Spiked matrix plates (10 x 20 cm) were prepared by pipetting 1 mL of a 1 or 10 $\mu g \ mL^{-1}$ pesticide calibration standard over the plate to yield surface area concentrations of 0.005 or 0.05 $\mu g \ cm^{-2}$, respectively. A blank plate was also extracted after spiking with 1 mL of acetone.

Pesticides were quantified by gas chromatography using external calibration standards purchased from ChemService (Westchester, PA). Compounds were separated on a Varian Model 3400 GC with a 30 m x 0.25 mm id capillary column containing 0.25 μm of SE54 (AllTech Assoc.). Helium flow rate was initially 1.9 mL min⁻¹. On-column injection was via autosampler at 60°C with a 0.1 minute hold and a ramp to 250°C at 50°C/minute. For chlorothalonil and the α and β isomers of endosulfan, the column temperature program consisted of an initial 80°C for 2.5 min with a ramp to 200°C at 10°C min⁻¹. A pulsed electron capture detector (ECD) was used for detection of chlorothalonil and endosulfan.

A statistically based limit of detection (LOD) and quantitation (LOQ) was determined for each pesticide based on the average recovery from the silica gel matrix at the estimated level of instrumental detection. The LOD, expressed as μg pesticide per cm² was defined as the average surface area concentration recovered ± 3 times the standard deviation (s.d.); similarly, the LOQ was defined as the mean recovery ± 6 s.d. (Table 1). The extraction efficiencies from silica gel plates were determined by a single analyst using 10 replicate samples fortified at different levels with calibration solutions (Table 1).

The stability of chlorothalonil and endosulfan on silica gel plates stored for up to 6 months was also determined; no significant decline in residue recovery was observed.

Table 1. Limits of Detection (LOD), Quantitation (LOQ), and Extraction Efficiencies (n=10) for Chlorothalonil and Endosulfan on Silica Gel Plates

	LOD	*LOQ*	*Extraction Efficiency @ Spiking Rate (μg cm⁻²)*	
Pesticide	*μg cm⁻²*	*μg cm⁻²*	*0.00005*	*0.005*
chlorothalonil	0.00010	0.00012	146±15	97±13
a-endosulfan	0.00006	0.00008	86±12	98±18
b-endosulfan	0.00005	0.00006	67±10	93±20

Water Monitoring

Water was sampled at selected locations shown on Figures 1 and 2 both before pesticide application and within an hour after application. A 1-L water sample was collected by immersing a bottle under the surface until the volume was filled. Water was also sampled at several intervals during a maximum of 6 h after application and again several times over the next 1-2 months. Samples were transported on ice to the Washington State Department of Agriculture Chemical and Hop Laboratory in Yakima, WA for analysis. The analytical procedures followed USEPA Method 525.1. Positive results were validated using a Varian Saturn 3 GC-Ion Trap MS. Appropriate field and laboratory matrix spikes were used and procedures carried out following standard operating procedures developed under FIFRA GLP (Federal Insecticide Fungicide & Rodenticide Act--Good Laboratory Practices) guidelines.

AgDRIFT Modeling

Deposited residues were graphed relative to meters from the edge of the stream or pond banks. AgDRIFT (ver. 2.03) model simulations were made using the Tier III forestry module, and the output was overlain on the deposition graphs. Input parameters to Tier III closely matched the actual spray application parameter with two exceptions. Nozzle size in AgDrift was limited to a D6 (core 46) that would give a smaller volume median diameter droplet size than the D8 nozzle actually used. Relative humidity was not measured at Site 1 so it was set to the the average R.H. during spraying at Site 2 (96%), although the spray runs had occurred about 1.5 h later in the day. Because humidity in western WA drops significantly between early morning and noon when it is not raining, an additional AgDRIFT simulation was made using 75% R.H. to reflect the later application time at Site 1. For simulations at site 2, R.H. was set to the highest recorded level (98%).

Results and Discussion

Many drift studies rely on controlled experimental applications that permit accurate demarcation of the spray swath and crosswind distances (24). Those studies also tend to monitor drift from only one or a few spray swaths. In this study, monitoring was conducted conjointly with commercial applications to two Christmas tree plantations. Each plantation received multiple swaths as deemed appropriate for the size of the treated plantation. Because the spray swath closest to the intended buffer zones could not be pinpointed apriori, monitoring plates were generally placed starting at the edge of the aquatic system and at various distances in a nearly perpendicular line to the estimated direction of spraying. At each site the highest rates of chlorothalonil recovery were assumed to be within a spray swath because the buffer width was only 15.2 m. The highest recoveries of endosulfan should have been adjacent to a swath because the buffer zone was supposed to be at least 92 m from the edge of the aquatic system. At Site 2 along transect 1, however, some of the plates should have been directly oversprayed.

Pesticide Deposition at Site 1

Chlorothalonil residues along transects 1 and 2 were highest on the slica gel plates placed 60 m and 30 m, respectively, from the stream bank. Because recoveries generally ranged from 40-70% of the theoretical application rate, the monitoring plates were likely placed inside the spray zone. The lowest recovery in the putative swaths corresponded to 0.2% (0.03 μg cm^{-2}) of the expected application rate (Figure 3, transect 1), and the highest recovery was 69% (16 μg cm^{-2}) of expected (Figure 3, Transect 2). The low recovery of chlorothalonil at the plantation end of transect 1 (91-m) suggests the plate may have been located between swaths. Althernatively, nearby trees may have deflected spray droplets from impinging on the plate. Large variations in percentage recovery of calculated application rates are consistent with other studies where terrain was not as variable as at the Christmas tree plantations. For example, deposition of endosulfan within a swath applied to a potato field bordered by grain field or fallow ranged from 70-170% of theoretical rates (*13*). When chlorothalonil was applied directly to a pond, which is ostensibly a very uniform surface, 67-88% of the theoretical application rate was recovered (*25*).

At Site 1, recovery of deposited chlorothalonil declined at least 10 fold at a distance of 5-10 m beyond the putative swath (Figure 3). Most of transect 1 was actually in a wooded area, but much of transect 2 was under the chlorothalonil swath with only two monitoring plates actually set in the riparian zone. Recovery of chlorothalonil in the wooded areas was generally <0.01% of the theoretical application rate; however, 0.04% (0.01 μg cm^{-2}) was recovered at the stream bank along transect 1.

Along both transects at Site 1, endosulfan was recovered at levels <0.1% of the theoretical application rate (Figure 4). Thus, endosulfan drifted into the monitoring area beyond the outer edge of its buffer zone. Along transect 1 and 2, 0.007% (0.0002 μg cm^{-2}) of the theoretical application rate was recovered at the stream bank (Figure 4). Despite the detections of pesticide residues at the stream bank, neither chlorothalonil or endosulfan were detected in water samples collected from Duffy Creek downstream of the plantation over a 24 h period post application.

Pesticide Deposition at Site 2

Transect 1 at site 2 monitored sprays parallel to the direction of the silica gel plates, and the distal 30 m should have been under a spray swath. At this site, plates were placed on posts along an open service path, eliminating the possibility of vegetation filtering out spray droplets. At distances of 76-107 m from the transect origin, chlorothalonil and endosulfan recoveries ranged from

13-17% of theoretical application rates (Figure 5, 6). The low recoveries suggested that the plates may have actually been at the edge of a swath. At distances less than 76 m from the origin of the no-spray buffer, recoveries of both pesticides declined by nearly 100-fold within 10 m (Figure 5, 6). Less than 0.01% of the applied active ingredient was recovered at the origin of the transect.

Transect 2 at site 2 ran nearly perpendicular to the direction of the spray swaths. Recoveries of chlorothalonil were higher from plates situated on posts than from plates on the ground, suggesting that vegetation filtered out some of the spray (Figures 5, 6). Recoveries from the ground plates placed outside of the buffer zone ranged from 9%-148% of the nominal application rate. Residues on post-mounted plates ranged from 65%-248% of the nominal application rate. In contrast to chlorothalonil deposition, endosulfan deposition on ground plates was higher than on post-mounted plates at distances of 76-107 m from the pond. This difference in recovery suggests that endosulfan was drifting onto the plates at a low angle (i.e., less than the 1.7-m post height) from its swath just beyond the monitored area.

Less than 0.001% (i.e., <LOQ) of the applied chlorothalonil was recovered in the 15-m buffered area around the pond, suggesting the residues moved beyond the buffer zone (Figure 5). The data could be interpreted as indicating drift of chlorothalonil over the pond in an inversion. Winds were calm and humidity was 98% at the time of spraying, so this possibility is plausible. One hour after application, 0.76 ppb of chlorothalonil was detected in one water sample collected near the edge of the pond but samples taken 3 and 29 hours later were negative. Endosulfan was not detected in any water sample.

AgDRIFT Simulation of Deposition

AgDRIFT is a simulation model that was designed for estimating near-field spray drift from ground and aerial applications (26). The computation module for aerial applications, AGDISP (AGricultural DISpersal), is a mechanistic type of model because it accounts for physical processes like gravitational acceleration, air resistance, droplet evaporation, and mode of application. AgDRIFT can be useful for determining no-spray buffer widths within 800 m downwind of an aerial swath. The aerial simulations can be made for either agricultural or forestry land, and a series of three tiers allows user input of an increasing number of application and meteorological variables.

Because the first version of AgDRIFT was only made publicly available in 1998, few field studies have been conducted to validate it. Recently, however, 161 field trials of aerial agricultural applications showed the model robustly predicted spray deposition within approximately 100-150 meters perpendicular to a spray swath (27). Experiments that were used for model validation involved only four flight lines and the fields were comparatively level with mowed grass

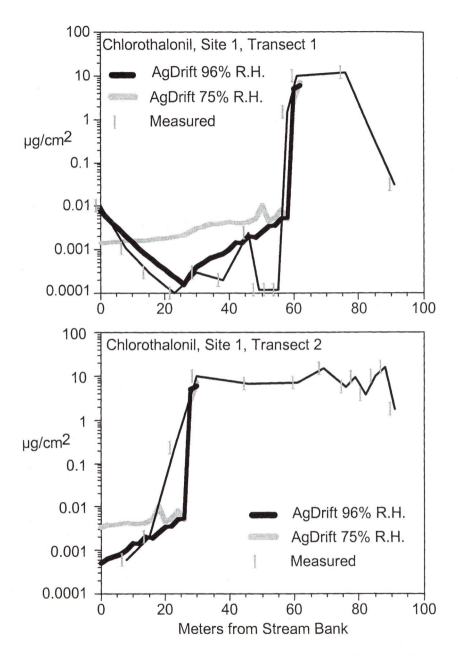

Figure 3. Measured and simulated deposition of chlorothalonil at Site 1 along two transects running from the stream edge into the sprayed plantation.

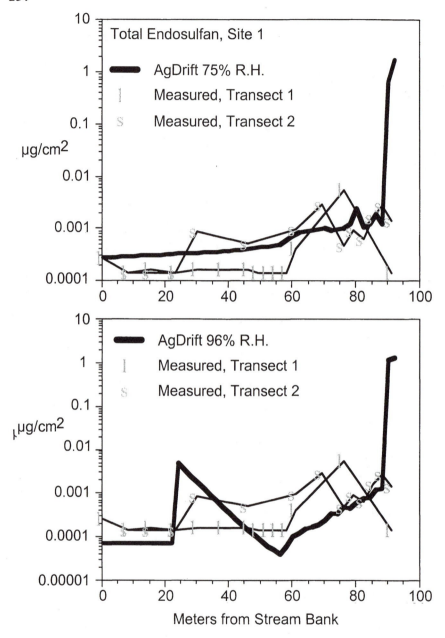

Figure 4. Measured and simulated deposition of endosulfan ($\alpha + \beta$ isomers) at Site 1 along two transects running from the stream edge into the sprayed plantation.

or disk and plowed, bare soil or sorghum stubble (28). Furthermore only five of the 161 field trials involved helicopter applications.

Our studies in Christmas tree plantations offered an opportunity to further validate AgDRIFT under more complex topography and using helicopter application. The resulting simulation output was placed on the same graph as the empirical coordinates (Figures 3-6), but the downfield drift distances were offset to correspond with the edge of the swath. Because R.H. was not measured at Site 1, two values, 75% and 96% were used as meteorological input parameters. One of the previously noted limitations of AgDRIFT is its sensitivity to evaporative effects (27), and thus R.H. may influence simulated downwind deposition. At an R.H. of 75%, AgDRIFT simulated

deposition of chlorothalonil matched measured deposition reasonably well beyond the spray swath (Figure 5). The upswing in residues recovered from transect 1 of Site 1 were also faithfully modeled by AgDRIFT. When R.H. was changed to 96%, AgDRIFT over predicted deposition by almost 10-fold along transect 2 and most of transect 1 (with the exception of an under prediction at the stream bank).

AgDRIFT with input parameters of 75% and 96% R.H. simulated endosulfan deposition reasonably well along both transects (Figure 4). Deviations of simulated residues from measured residues along much of the transect were 10-fold or less. The actual residues ranged from < LOQ to a maximum of 0.08% of the theoretical deposition rate so error in simulation of these low levels is expected. Notably, the non-monotonic variability in recovery of endosulfan residues along the transect was also simulated by the model. Thus, measured residues at 46 m were notably higher than residues at 49 meters, and similarly, simulated residues at approximately 50 m were higher than residues at 54 m.

Model simulations at Site 2 over predicted chlorothalonil and endosulfan residues by a factor of about 10-fold and 40-fold, respectively, at the 0-m deposition point along transect 1 (Figure 5, 6). Humidity at initiation of spraying was 98%, but the significantly lower recovery of measured residues suggests that evaporation under field conditions may be much faster than predicted by the algorithm driving AgDRIFT. This difference between modeled deposition and field deposition at increasing distances was noted in the validation studies of AgDRIFT and attributed to the sensitivity of AgDRIFT to wet bulb depression (Bird 2002).

Notably, the slope of deposition predicted by AgDRIFT between 30 and 70 m mirrored closely the slope of the actual deposition curve along transect 1 at Site 2. On the other hand, AgDRIFT failed to adequately simulate deposition of endosulfan along most of transect 2 (Figure 6). Part of the problem could have been a direct overspray along transect 2 if the first spray swath had started inside of the 91.5 m no-spray buffer zone. However, deposition of endosulfan within the riparian management zone (0-15 m) was simulated accurately; i.e., AgDRIFT predicted no deposition of residues in this location.

In addition to the sensitivity of AgDRIFT to humidity, another source of modeling error may have been the unavailability in the AgDRIFT nozzle library

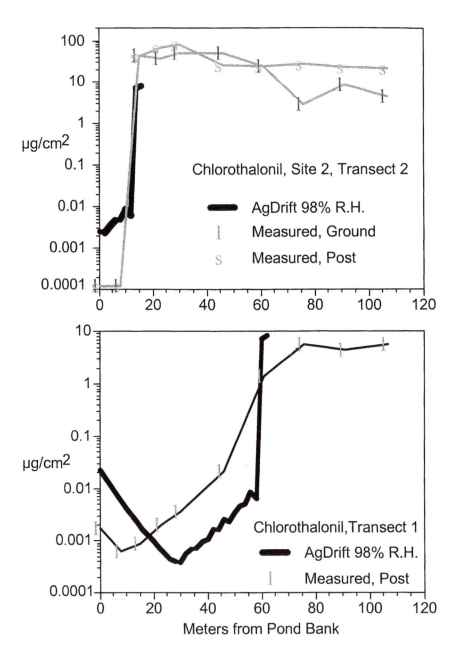

Figure 5. Measured and simulated deposition of chlorothalonil at Site 2 along two perpendicular transects originating in the no-spray buffer zone.

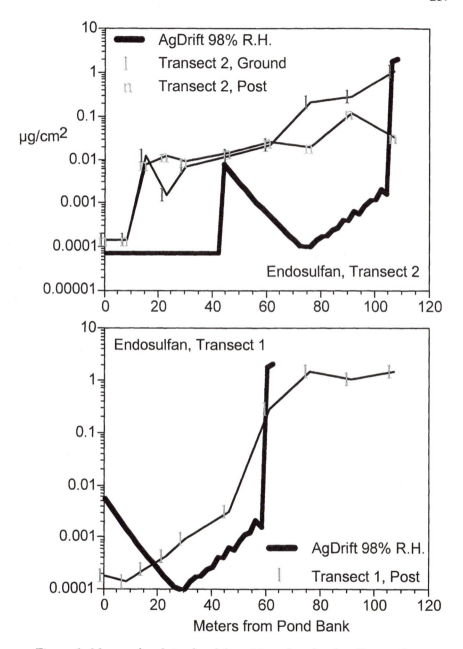

Figure 6. Measured and simulated deposition of total endosulfan residues at Site 2 along two perpendicular transects.

of the actually used nozzle type. The D8 (core 46) nozzle would produce droplets with a larger volume median diameter thus hypothetically less drift than the D6 (core 46) nozzle that AgDRIFT defaulted to.

Conclusions

Riparian management zones that also serve as part of a no-spray buffer zone will filter out pesticide residues. The pesticide residues can be detected at the water edge, but the levels depositing in a flowing stream or large pond are probably rapidly diluted to concentrations that are too low to accurately detect. In this study, only one transient detection of endosulfan was noted in the pond at Site 2, but detection could not be confirmed with additional detections in water samples collected subsequently. Thus, the buffer width sizes required by Federal regulations for endosulfan and required in 1994 under the Washington State Forest Practice Rules for aerial applications of chlorothalonil seem protective of aquatic resources.

Pesticide deposition could be reasonably simulated with the forestry tier of AgDRIFT, but sometimes residues farthest from the spray swath were significantly overestimated. The aerial deposition model of AgDRIFT is very sensitive to humidity. Proper simulation of field results will thus depend on input of appropriate meteorological factors.

Buffer zones are usually set up to prevent direct movement of pesticide residues into aquatic systems. Best management practices for spray operations are recommended to minimize deposition within the inner areas of the riparian management zone. Pesticide spray drift deposition studies to further validate AgDRIFT simulations will facilitate assessments of the risk of adverse effects within riparian zones when they overlap with no-spray buffer zones.

Acknowledgements

This research is a contribution from the Washington State University College of Agriculture & Home Economics and was funded in part by grants from the USDA and the Washington State Department of Agriculture. C. Graber, D. Davis (WA State Department of Ecology), and T. Mallgren (WA State Department of Natural Resources) provided technical assistance.

References

1. Muscutt, A. D.; Harris, G. L.; Baily, S. W.; Davies, D. B. *Agric. Ecosys. Environ.* **1993**, *45*, 59-77.

2. Arora, K. S.; Mickelson, K.; Baker, J. L.; Tierney, D. P.; Peters, C.J. 1996. Herbicide retention by vegetative buffer strips from runoff under natural rainfall. *Trans. Am. Soc. Ag. Eng.* **1996**, *39*, 2155-2162.

3. Nowell, L. H.; Resek, E.A. *Rev. Environ. Contam. Toxciol.* 1994, **140**, 1-164.

4. Wagner, R. J.; Ebbert, J. C.; Roberts, L. M.; Ryker, S. J. U. S. Geological Survey Water-Resources Investigations Report 95-4285, Tacoma, WA. **1996**, 52 pp.

5. U.S. EPA. *Draft pesticide registration notice: spray and dust drift label statements for pesticide products.* Office of Pesticide Programs Pesticide Registration (PR) Notice 2001-X. **2001**, http://www.epa.gov/opppmsd1/PR_Notices/prdraft-spraydrift801.

6. U.S. EPA. *EFED risk assessment for the reregistration eligibility decision on endosulfan (Thiodan).* **2001**, http://www.epa.gov/

7. Washington State Administrative Code (WAC) 222-38. *Forest Chemicals.* **2001**, http:www.wa.gov/dnr/htdocs/fp/fpb/fprules2001/wac222-38.pdf.

8. Washington State Department of Natural Resources. *Washington Forests and Fish Report.* **1999**, http://www.wa.gov/dnr/htdocs/fp/fpb/forests&fish.html#APPE

9. Marrs, R. H.; Frost, A. J.; Plant, R. A.; Lunnis, P. 1992. *Biol. Conserv.* **1992**, 59, 19-23.

10. Marrs, R. H.; Frost, A. J.; Plant, R. A.; Lunnis, P. 1993. Determination of buffer zones to protect seedlings of non-target plants from the effects of glyphosate spray drift. *Agric. Ecosys. Environ.* **1993**, *45*, 283-293.

11. Payne, N. J.; Feng, J. C.; Reynolds; P. E. *Pestic. Sci.* **1990**, *30*, 183-198.

12. Payne, N. J. *Pestic. Sci.* **1992**, *34*, 1-8.

13. Ernst, W. R.; Jonah, P.; Doe; K.; Julien, G.; Hennigar, P.. *Environ. Toxicol. Chem.* **1991**, *10*, 103-114.

14. Davis, B. N. K.; Lakhani, K. H.; Yates, T. J.; Frost, A. J; Plant,R. A. *Agric. Ecosys. Environ* **1993**, *43*, 93-108.

15. Davis, B. N. K.; Brown, M. J.; Frost, A. J.; Yates, T. J.; Plant., R. A. *Ecotox. Environ. Safe.* **1994**, *27*, 281-293.

16. Wilson, A. G. L.; Harper, L. A.; Baker, H.. *Austr. J. Exp. Agric.* **1986**, *26*, 237-43.

17. Payne, N., Helson; B. Sundaram, K;. Kingsbury, P.; Fleming, R; de Groot, P. *Estimating the buffer required around water during permethrin applications.* Information Report FPM-X-70, Forest Pest Management Institute, Canadian Forestry Service, Sault Ste. Marie, Ontario, Canada. **1986**, 26 pp.

18. Payne, N. J.; Helson, B. V.; Sundaram, K. M. S.; Fleming, R. A. *Pestic. Sci* **1988**, *24*, 147-161.

19. Feng, J. C.; Thompson, D. G.; Reynolds, P. E. *J. Agric. Food Chem.* **1990**, *38*, 1110-1118.

20. Rashin, E., Graber, C. *Effectiveness of best management practices for aerial application of forest pesticides.* TFW-WQ1-93-001, Washington State

260

Department of Ecology, Watershed Assessments Section, Olympia Washington. **1993**, 86 pp.

21. Teske, M. E.; Bird, S. L.; Esterly, D. M.; Ray, S. L.; Perry, S. G.. *A User's Guide for AgDRIFT 2.03: A Tiered Approach for the Assessment of Spray Drift of Pesticides.* C. D. I. Report No. 01-01, Steward Agricultural Research Services, Inc., Macon, Missouri. **2001**.

22. Elling, W.; Huber, S. J.; Bankstahl, B.; Hock, B. 1987. Atmospheric transport of atrazine: a simple device for its detection. *Environ. Pollut.* **1987**, *48*, 77-82.

23. Hill, B. D.; Inaba, D. J. *J. Enviro. Sci. Hlth.* **1990**, *B25*, 415-432.

24. Bird, S. L. In *Agrochemical Environmental Fate*; Leng, M. L.; Leovey, E. M. K.; Zubkoff, P. L., Eds., CRC Lewis Publishers, Boca Raton, FL. **1995**, pp. 195-207.

25. Ernst, W.; Doe, K.; Jonah, P.; Young, J.; Julien, G.; Hennigar, P. *Arch. Environ. Contam. Toxicol.* **1991**, 21, 1-9.

26. Teske, M. E.; Bird, S. L.; Esterly, D. M. ; Curbishley, T. B.; Ray, S. L.; Perry , S. G.. *Environ. Toxicol. Chem.* **2002**, *21*, 659-671.

27. Bird, S. L.; Perry, S. G.; Ray, S. L.; Teske, M. E.. *Environ. Toxicol. Chem.* **2002**, *21*, 672-681.

28. Hewitt, A. J.; Johnson, D. R.; Fish, J. D.; Hermansky, C. G.; Valcore, D. L.. *Environ. Toxicol. Chem.* **2002**, *21*, 648-658.

Chapter 15

Pesticide Use and World Food Production: Risks and Benefits

Gerald R. Stephenson

Department of Environmental Biology, University of Guelph, Guelph, Ontario N1G 2W1, Canada

For most of the next century, we will need to produce enough food for nine billion people instead of the six billion we are trying to feed today. If we try to improve the average diet as well, we may need to double annual world food production for most years of the next century. There is not much more land that can be devoted to agriculture without having an enormous environmental impact on forested or wilderness areas. Furthermore, a higher proportion of agricultural land may be used industrially to produce fuel or fibre instead of food. Thus, we may need to grow twice as much food on even less land than we are using today. We are currently using $35 billion worth of pesticides each year in agriculture, world wide. What will the benefits and risks be if this level of pesticide use is continued or increased? What will they be if pesticide use is discontinued? Several years ago, farmers in highly developed, industrialized countries could expect a three or four fold return on money spent on pesticides. Is this still true? Can we meet world food demands if producers stop using pesticides because of reduced economic benefits? Can better IPM preserve the economic benefits of pesticide use? Although crop losses are currently greatest in less industrialized countries, can we meet the educational and training requirements to safely increase pesticide use in these areas? These are just some of the questions facing scientists and pest management experts as agriculture faces its greatest challenge in history between now and the year 2100.

Introduction

The pesticide industry is very big business. World-wide pesticide sales now exceed $35 billion per year (*1*). Sales have exceeded $20 billion per year since the 1980's and herbicides account for at least half of the business. More than half of the world's pesticides are used in Europe and North America., 25% are used in the far east and approximately 25% in the rest of the world combined. Latin America, particularly Brazil, is an area where there is perhaps the greatest potential for the pesticide market to expand. South America is also the part of the world where there is the most "new land" that could sustain use for agriculture. At present, developing countries in warmer climates use half of the insecticides whereas industrialized countries in more temperate climates use most of the herbicides. For example, herbicide use accounts for 70% and 80% of total pesticide use in the USA and Canada, respectively.

Health Risks Associated with Pesticide Use

Unfortunately, pesticide misuse can be a human health risk. In his recent book, Don Echobichon (*2*) estimates that there are numerous accidental deaths and thousands of accidental pesticide poisonings in North America each year. In developing countries, there are millions of reported poisonings and hundreds of thousands of pesticide related deaths each year. How many incidents go unreported? What about the chronic health effects of pesticide use in developing countries? Fortunately, the health risks associated with pesticide use are largely a preventable problem. In fact, it is accurate to say that 'proper' pesticide use rarely results in a human health problem. The industry has a major role in preventing human health effects with their continuing efforts to develop even safer pesticides. However, safe pesticide use requires safe equipment and good systems for training and educating pesticide applicators and farm workers. Good regulations of pesticide sale and use that are regularly enforced are also essential. It is costly to maintain these systems in industrialized countries. It is a major challenge to get adequate regulatory systems established in developing countries.

Environmental Risks Associated with Pesticide Use

Pesticide use can also present risks to the environment. To visualize the factors that contribute to this risk, it may be helpful to think of the following equation or model,

$$ER = f(V, P, M, T)$$

Quite simply, the environmental risk (ER) of any chemical depends on the volume used (V), the persistence of the chemical (P), its mobility in the environment (M) and its potential toxicity (T) to non-target organisms. We now have regulatory systems to prevent the introduction of persistent, bio-accumulating pesticides like DDT for use in agricultural environments. Our early experience in managing spray drift and vapor drift with 2,4-D is helping us prevent similar problems with newer herbicides like glyphosate and clomazone. Likewise, our earlier experience with triazine carry-over, soil residue problems are a guide in managing similar problems with the sulfonyl urea, imidazolinone and other new low-rate, soil active herbicides.

With few exceptions, our experience with older pesticides has been quite good. For example, in a study conducted at Rothemsted in England (3), research plots were treated with at least five pesticides each year for 20 years. Seventeen months after pesticide use was discontinued, there were no detectable pesticide residues in the plots. Furthermore, there were no differences between treated and control plots with respect to either soil microbial processes or the yield of barley used as an indicator crop. The increasing use of the new "low-rate" pesticides is certainly reducing the risks for non-target organisms in the environment as a whole. However, our farmers are struggling with the management of "on farm risks" associated with the "carry-over" residues of the new "low-rate" herbicides in soil and their potential for injury to subsequent crops grown in rotation. In Canada, the labels for many of these products require farmers to conduct their own field bioassays to assess the safety for candidate rotation crops. Such requirements seem unfair and unmanageable.

Evaluating the Benefits of Pesticide Use

For our consideration of pesticide benefits, we should ask, benefits to whom? At the 1998 IUPAC meetings in London, England, Sir Colin Spedding (4) encouraged us to consider benefits to manufacturers, growers, processors, and consumers or citizens. Of course there are large differences in the numbers of people in these various groups. For example, there are relatively few people associated with the manufacture of pesticides, while the number of consumers or citizens impacted by pesticide use is enormous. Furthermore, in industrialized countries, there is a low percentage of people who would be involved with pesticide application whereas in developing countries, the application of pesticides with hand operated equipment may involve a larger number of people.

Manufacturers

In the development of a new pesticide, the financial benefits and risks to the manufacturer are huge. At 8 or 10 years after discovery, when the first product is sold, investments in research and development may total nearly $100 million (5) Companies need to know that a candidate chemical far exceeds current government requirements for health and environmental safety, very early in development, to be sure that the chemical will have a sustainable life on the market. The current trend toward fewer but larger companies simply reflects the reality, that only a very large company can manage the risks associated with the huge investments required to develop new pesticide products.

Growers

What about benefits to the grower? For many years, we have commonly assumed that there is about a $4 return to the grower for every dollar spent on pesticides. Is this still true? Most studies indicate (6)about a 30% yield benefit when pesticides are used. For the USA in 1997, when a 30% increase in crop value was compared to total expenditures on pesticides, the return was approximately $3-$4 for every dollar spent on pesticides.

Several industrialized countries are in the midst of programs to reduce the use of pesticides in agriculture. Canada is one of the leading countries in this endeavor. In the Province of Ontario, there is a program called Food Systems 2002 which was initiated as a political campaign promise during the 1980's (7) The goals of this program are two-fold; (1) To reduce pesticide use in agriculture by 50% by the year, 2002; and (2) To accomplish this without reducing agricultural productivity. In a sense, the goal was to increase grower education and to develop better pest monitoring and effective alternatives to chemical pesticides so that there could be nearly a 100% reduction in the use of pesticides when they were not needed. According to the most recent survey of pesticide use in 1998 (8), total kg of agricultural pesticide use in Ontario was reduced by nearly 40% compared to 1983. In fact, the reduction was nearly 50% in maize (corn) where herbicides are the predominant pesticides used. Of course this is largely due to a shift to new low-rate herbicides in corn. Environmental critics complain that the shift to lower rate, more powerful pesticides may not be reducing environmental risks. To evaluate these concerns, Gallivan et. al. (9) employed the "Kovach model" for calculating the Environmental Impact Quotient (EIQ) for pesticide use. This model includes estimates for farm worker risk, consumer risk and ecological risk. Because of data gaps for some pesticides and other problems, it does have its short comings. However, it is one of the best current methods for comparing the impact of one pesticide versus another or pesticide use in one year versus another year. When pesticide use in Ontario was examined with this model, it was apparent that the overall EIQ per hectare in 1998 was only 66% of what it was in 1983, a 34% reduction. Furthermore, because of higher crop yields, the EIQ per tonne of crop produced in 1998 was 58% of what it was in 1983, a 42% reduction. This is a definite benefit to the citizens of Ontario and to their environment. However, what about economic returns on pesticide use to the growers. A recent study by Teague and Brorsen (1995) indicated that for the ten major agricultural states in the USA, returns to the grower on pesticide use declined from $8 per dollar in 1949 to approximately $4 per dollar in 1991. One would expect that the success of the Food Systems 2002 program in Ontario, with respect to preventing un-needed pesticide use, might increase the economic returns to growers on money spent on pestaicides. This proved not to be true. When we assumed the commonly accepted, 30% yield benefit with pesticide use and compared increased crop values with total expenditures for pesticides, Ontario growers had a return of $3.22 per dollar spent on pesticides in 1998 compared to $8.42 per dollar in 1983. The main reason for this was that pesticide expenditures had increased 8-fold since 1973 with only a 3-fold increase in crop value,despite dramatic increases in crop yields. . The conclusions are obvious. It probably costs manufacturers at least six times as much to develop a new pesticide today as it did in the early 1970's. Higher pesticide prices today reflect this. However, there has not been a parallel increase in the value of agricultural commodities. Thus, in industrialized countries, programs to eliminate pesticide use when and where there is little chance for a yield benefit not only reduce potential risks to the environment, they are absolutely essential to preserve economic benefits of pesticide use for the growers.

Society

What about the economic benefits of pesticide use to society as a whole? Pimentel and Greiner (11), at Cornell University, estimate that the $6.5 billion spent on pesticides by farmers in the USA prevented about $26 billion in crop losses due to pests, again about a $4 return to the growers for each dollar spent on pesticides. However, they pointed out that we should also consider the $8 billion in indirect costs to society for regulating, preventing and correcting environmental and health problems associated with pesticide use. With this approach, the net economic benefit to society in general was closer to $2 per dollar spent on pesticides.

Pesticides Save Human Labour

Pesticide use, particularly herbicide use for weed control, reduces hand labour requirements for agriculture. Ontario agriculture probably reflects the norm in industrialized countries and only 2% of our population is involved in production agriculture. In other words, one person can produce enough food for 50 other people. In developing countries, this is far from true. World-wide, 46% of the population is involved in field work for agriculture. In Brazil it is 20%; Mexico, 25%; and in Kenya it is 70% or two people in every three (*12*). In too many parts of the world, too many people, especially women and children, are deprived of an education and chances for a better standard of living because their labor is needed to weed and to harvest crops.

Pesticide use saves energy

It is often assumed , that as agriculture has become more intensive and more dependent on technology - energy requirements have increased. In fact, energy requirements for crop production in Ontario did increase between the 1940's and the 1970's, largely due to a 1000 % increase in energy requirements for producing nitrogen fertilizer (*13*)However, these trends are now reversed. More efficient methods have reduced energy requirements in fertilizer production by at least 40%. Swanton et.al.(*14*) have shown that herbicide use increases the energy efficiency in both corn and soybean production. This is largely due to eliminating the need for primary tillage (plowing). Furthermore, in crops like soybeans, energy efficiency is even greater if at least one secondary tillage operation (rotary hoeing) can be eliminated. The trend toward the new, low-rate herbicides is also decreasing the energy investment in each herbicide application. As fuel costs continue to increase, the energy benefits of pesticide use should continue to increase as well.

World Food Production

Oerke et.al. (*15*), have estimated that the use of crop protection chemicals doubled the yields of the world's eight principal cash crops between 1965 and 1990. For the agricultural land involved in 1990, there was a potential to produce $579 billion in food, world-wide. They estimated that pesticide use in agriculture doubled yields from 30% of the potential without pesticides to 60% of the theoretical potential. However, pests were still causing an approximate 40% loss in total food production. Successes with pest control in agriculture, varied with the crop and the regions in the world where the crop was grown. Losses due to pests in maize (corn) were less than 30% in Europe but greater than 50% in Africa. Losses in wheat production were less than 30% in Europe but greater than 40% in what was the former USSR. Losses in rice production were less than 30% in Oceania but greater than 50% in Africa and the Americas. Overall, crop losses due to pests in Africa were double what they were in Europe. However, even in most developing countries, food production is increasing faster than the increase in population. Thus, food production per capita is increasing. Despite an increasing world population, the actual number of malnourished people is decreasing (5). That is good news. However, the bad news is that there are still more than one half billion undernourished people in the world and in Sub- subsaharan Africa, the number is still increasing (5).

World Land Use

The potential impact of pesticide use on the environment is discussed throughout the world. However, what about the impact of 'agriculture' on the environment? If we were still producing crops with the yields of 1960, we would need nearly three times as much land for agriculture - an area equal to all of the land currently used for agriculture in Brazil, Europe and the USA, combined (16).Lester Brown (17) estimates that if farmers did not have the fertilizers and pesticides to "nearly triple land productivity since 1950, it would have been necessary to clear half of the world's forested land for food production. If this technology and other technology is not available in the future, how much more land will be consumed by agriculture and lost for other uses for our children and grandchildren?

World Population Trends and Needs for the Next Century

As recently as the late 1960's, Paul Ehrlich was warning us with his book, "The Population Bomb", that we would already be experiencing a world-wide catastrophe because of over population. Fortunately, this has not happened. The rate of population increase has already peaked and is beginning to decline. In every part of the world except Africa, populations are aging. Current, conservative estimates (UN and others) are that our present population of 6.24 billion (18) will continue to increase for only another 50 years and will peak at close to 9 billion in about 2050 (17). Beyond 2050, world population should begin to decline. Keep in mind, that these are the best possible predictions. However, it is only recently that we could begin think in such positive terms about world population trends. What this means for most of the 21st century is the following:

- 50% more people will need food, 9 billion instead of 6 billion
- a higher standard of living for people in developing countries could mean 50% more buying power for food
- this could mean that people in developing countries will consume less rice and sorghum, more wheat and maize, more potatoes and vegetables, more fruits, more dairy products, more animal protein.

Thompson (19)

This would mean that although food production has already been doubled or tripled in the last 50 years, we need to double it again. However, there is hardly 10% more rain-fed, arable land that would be sustainable for use in agriculture. In addition, more land may be diverted from food production to the production of fuel or fibre. We may decrease world food losses 30 to 40% by further preventing losses due to pests in the field and in storage (15). However, we need to double world food production again by about the year, 2025, on about the same land that we are using now (19). A 40% increase in food available for the world should be possible with better use the technology that we already have. However, a 100% increase in world food production will require new advances not only in pest management and pesticide technology but in other technology such as crop genetics..

"The Mental Affluence Trap"

Can we meet these challenges, as we have met the earlier challenges for world agriculture? Technologically, it is quite possible. Psychologically and culturally, it is much less certain. Particularly worrying, are the changing attitudes among the more affluent people of the world. It is what Hans Mohr (20) of the Universitat, Freiburg calls the "The Mental Affluence Trap". According to him, the willingness of people to accept change (new technology) is inversely proportional to their affluence. This attitude eventually leads to a mental immobility among the more affluent members of society, who become more and more critical of the advances and technology that were originally responsible for their prosperity. Conversely, less affluent people will more readily accept the potential risks of change in attempts to improve their prosperity. Therefore, people in developing countries will likely favor increased pesticide use to improve their health, whereas the more affluent people in industrialized countries will want to decrease pesticide use and consume more organic food in an attempt to preserve the good health that they already enjoy.

Summary and Conclusions

- Pesticide use has had a major role in tripling world food production during the last 50 years.
- Pesticide use benefits humans and their environment by reducing world hunger and by saving human labour, fossil fuels, and land
- However, world food production must again be doubled for most years of this century.
- Pesticide use as well as other technology will be essential to prevent the encroachment of agriculture onto unsuitable land, even wilderness land, that would not be sustainable for agricultural use.

- Efforts to reduce the use of agricultural pesticides where and when there is little chance to improve food production should continue. Such efforts minimize environmental risks and maximize economic benefits associated with pesticide use.
- A wide-scale reduction in "needed" pesticide use for agriculture in industrialized could make it more difficult to meet world food needs.
- There will be pressures to increase pesticide use in developing countries. However, we must be sure that educational and regulatory needs are met to prevent adverse health and environmental effects.
- It is important to minimize the impact of pesticides on the environment. However, it is just as important to maximize agricultural efficiency in order to minimize the impact of agriculture on the world environment.
- Our goal for the next 100 years should be to prevent human hunger without irreversible harm to the world environment. We have a greater chance to achieve this goal with integrated pest management, including the use of pesticides and other technology than with a major shift to organic farming.
- If the world population peaks at 9 billion in 2050 and declines to about 5 billion in 2125, future generations may have the choice between wide scale dependence on organic farming or reducing the amount of land devoted to agriculture. It would be selfish, narrow minded and short sighted to think that we have these choices today.

References

1. Hopkins, W.L. *Global Herbicide Directory - 2nd Edition.* Ag. Chem. Information Services. Indianapolis, IN, 1997.
2. Echobichon, D.J. *Occupational Hazards of Pesticide Exposure.* Taylor and Francis, Philadelphia, PA. 1998.
3. Evans, D.A. In *Pesticide Chemistry and Bioscience - The Food Environment Challenge.* Brooks, G.T.; Roberts, T.R., Eds, Royal Society of Chemistry, Cambridge, UK.1999; pp 3-34.
4. Spedding, C. In *Pesticide Chemistry and Bioscience - The Food Environment Challenge.* Brooks, G.T.; Roberts, T.R.,; Eds. Royal Society of Chemistry, Cambridge, UK. 1999, pp 405-410.
5. Klassen, W. In Options 2000, Proc. Eighth Int. Congress Pesticide Chem., Ragsdale, N.N.; Kearney, P.C.; Plimmer, J.R.; Eds. Am. Chem. Soc., Washington, D.C. 1995. pp 1-32.
6. Fernandes-Cornejo, J.: Jans, S.; Smith. *Rev. Ag. Econ.* **1998**,20(2), 462-488.
7. Surgeoner, G.A.; Roberts, W.; In *The Pesticide Question: Environment, Economics and Ethics.* Pimentel, D.; Lehman, H.; Eds. Rutledge, Chapman and Hall, Inc., New York, 1993. pp 206-222.

8. Hunter, C.; McGee. *Survey of Pesticide Use in Ontario,* Ontario Ministry of Agriculture and Food and Rural Affairs. 1998. ISBN 0-7743-9959-7, Guelph, ON, Canada.
9. Kovach, J.; Petzoldt, C.; Degni, J.; Tette, J. *New York Food and Life Sci. Bull.* **1992**, 139, 1-8.
10. Teague, M.L.; Brorsen, B.W.; *J. Ag. Appl. Econ.* **1995**, 27, 276-282.
11. Pimentel, D.; 'Greiner, A.; In *Techniques of Reducing Pesticide Use: Economic and Environmental Benefits.* Pimentel, D.; Ed. John Wiley and Sons, New York, 1997. pp 51-78.
12. Akobundu, I.O.; Third IWSS Congress. **2000**. Abstract No. 4, p 2. Foz Du Iquassu, Brazil.
13. Commoner, B. *The Closing Circle.* Bantam, New York, 1972.
14. Swanton, C.J.; Murphy, S.D.; Hume, D.J.; Clements, D.R.; *Agricultural Systems.* **1996**, 52, 399-418.
15. Oerke, E.C.; Dehne, H.W.;, Schonbeck, F.; and Weber, A.; *Crop Protection and Crop Production - Estimated losses in Major Food and Cash Crops.* 1994. Elsivier.
16. Avery, D.T. *Saving the Planet with Pesticides and Plastic.* Hudson Institute, Indianopolis, IN. 1995.
17. Brown, L.B. *Eco-Economy.* W.W. Norton & Co. New York. 2001.
18. U.S. Census Bureau. www.census.gov/cgi-bin/ipc/popclockw July 7, 2002.
19. Thompson, R.L. Proc. 23rd Int. Conf. Ag. Econ. Ashgate, VT, USA. pp 1-17.
20. Mohr, H. In *Pesticide Chemistry - Advances in International Research, Development and Legislation.* Frehse, H.; Ed. VCH, Weinheim. 1991. pp 399-418.

Chapter 16

Probabilistic Assessment of Pesticide Risks to Birds

Andy Hart

Central Science Laboratory, Sand Hutton, York YO41 1LZ, United Kingdom

Probabilistic risk assessment methods offer substantial advantages for assessing the impacts of pesticides on birds. These were explored using a simple example, involving insectivorous birds exposed to chlorpyrifos in apple orchards. Unlike current approaches for assessing pesticide risks, probabilistic methods do not rely on subjective 'worst-case assumptions' and safety margins. Instead, probabilistic methods allow the risk assessor to quantify the effects of variability and uncertainty on risk. They estimate the probabilities that different levels of effect will occur. This provides a more realistic and meaningful measure of risk. However, substantial further work is required to implement probabilistic methods and achieve a consensus on how they should be used.

Introduction

Current methods for assessing pesticide risks are deterministic, in that they use single, fixed values to estimate toxicity and exposure, and produce a single measure of risk (e.g. a risk quotient or toxicity-exposure ratio). In the real world, toxicity and exposure are not fixed, but variable. Furthermore, many aspects of risk assessment involve uncertainty – for example, when extrapolating toxicity from test species to humans or wildlife. Consequently, the effects of pesticides are both variable and uncertain.

Deterministic methods cannot incorporate variability and uncertainty directly. Instead, uncertain or variable parameters are fixed to worst-case values, or dealt with by applying assessment factors (sometimes called safety factors or uncertainty factors) based on expert judgement, or simply ignored.

Probabilistic methods can incorporate variability and uncertainty directly, by using probability distributions instead of fixed values for uncertain or variable parameters. These distributions can then be combined, to estimate a distribution for the measure of risk. This provides a much more complete description of the range of risks, which can be very helpful for decision-making. For example, instead of producing a single value for the toxicity-exposure ratio, probabilistic methods can estimate how often the ratio will exceed a regulatory trigger.

Probabilistic methods have been developed over many years and are actively used in other fields such as engineering, insurance, finance, and chemical contamination. There is widespread interest in applying them to the ecological risks of pesticides, both in North America (1,2,3) and Europe (4).

This paper reports a case study exploring the application of probabilistic methods to assessing pesticide risks to birds. The aim of the case study was to explore some of the potential advantages and disadvantages of probabilistic methods. It was not intended to define how probabilistic methods should be used: this will require further work.

This case study was initiated in conjunction with the US Environmental Protection Agency's ECOFRAM project (1). It therefore focussed on methods proposed by ECOFRAM, but the approach and conclusions are of general relevance for the implementation of probabilistic risk assessment for pesticides.

Case study scope and scenario

The case study focussed on risks to birds from the use of the organophosphorus insecticide chlorpyrifos in apple orchards in the United Kingdom, applied by air-blast sprayer at 0.96 kg/ha. The focal species selected for the example was the blue tit (*Parus caeruleus*), a small insectivorous bird which is common in orchards in the UK.

The scope of the case study was limited to considering acute lethal effects, exposure via the dietary route only, and an exposure period of one day. These limitations were intended to provide a simple illustration of the approaches. A risk assessment for regulatory purposes should take account of additional effects and timescales.

Deterministic assessment

The study began with a deterministic assessment, conducted in a manner consistent with standard European regulatory practice. Exposure was estimated using the following equation, which is a simplified version of the calculation proposed by ECOFRAM (*1*):

One day dietary dose (mg chlorpyrifos/kg bodyweight) =
$$\frac{TFIR \times PT \times FDR \times C}{W} \tag{1}$$

where

$TFIR =$	Total Food Intake Rate (kg dry weight/day).
$PT =$	Proportion of food obtained from Treated area (unitless).
$FDR =$	Fresh to Dry weight Ratio (unitless).
$C =$	Concentration of chlorpyrifos on food (mg chlorpyrifos/kg wet weight of food).
$W =$	Weight of bird (kg).

As is normal in a preliminary assessment, it was assumed that birds obtain all their food in the treated area, so *PT* was set to 1. Blue tits are assumed to feed entirely on small insects, so the term *PD* that is used by ECOFRAM (*1*) for 'proportion of diet' can be set to 1 and is omitted from the equation. Body weight *W* was set to 13.3g, an average value for blue tits (*5*). Daily total food intake *TFIR* was estimated as 3.3g dry weight/day using a standard equation published by Nagy (*6*). Fresh to dry ratio *FDR* was set to an approximate value of 5, and is necessary to convert *TFIR* to wet weight. *C* was set to 27.8 mg chlorpyrifos/kg food (wet weight), based on the approach suggested by Kenaga (*7*) as applied in Europe (*8*) (i.e. assuming that initial residues on the small insects eaten by blue tits are similar to those on plant material of similar surface area / volume ratio, and taking the 'typical mean' value from Kenaga's analysis). These estimates resulted in an estimated one day dietary dose of 34.5 mg/kg.

The one day dose was compared to 32 mg/kg, the median lethal dose (LD50) from a study with the bobwhite quail, a standard test species (*9*). In Europe this comparison is done by calculating the ratio of toxicity (the LD50) to exposure (the daily dose). This results in a TER of 0.9. This is well below the threshold of 10 which is set down in EU regulations, so the pesticide could not be authorised for sale unless a more refined assessment showed the risk to be acceptable (*10*). In North America this comparison would be done by calculating the inverse ratio (i.e. exposure/toxicity), which is called the 'risk quotient', and would lead to a similar conclusion (further assessment required).

Probabilistic assessment

The probabilistic assessment explored some of the approaches proposed by ECOFRAM (*1*), in particular: (a) methods for extrapolating acute toxicity between species, (b) a method for using toxicity data to calculate the tolerances of individual birds, (c) the use of Monte Carlo simulations for propagating uncertainty, (d) progressive refinement of the assessment, by incorporating additional distributions in place of worst-case assumptions, (e) the use of generic field data to estimate distributions of pesticide residues on invertebrates, (f) the use of radio-tracking to estimate empirical distributions for the proportion of food obtained by birds from treated areas.

Progressive refinement was achieved by developing a series of three models. Model 1 used distributions for toxicity, but a worst case estimate for exposure. Models 2 and 3 introduced distributions for exposure by replacing fixed values with distributions, first for C and then PT. A summary of the data used in the three models is given in Table I.

The calculations were executed using the computer programs @Risk (Version 3.5.2 for Excel, ©Palisade Inc.) and Microsoft Excel 97 (©Microsoft Corporation).

Model 1 – a basic assessment

Model 1 differed from the deterministic assessment by (a) using a probabilistic approach for estimating toxicity and (b) generating a distribution for the percentage mortality of exposed individuals as output, rather than a fixed estimate of the toxicity-exposure ratio. Model 1 used the same method for estimating exposure as the deterministic assessment, described earlier, resulting in a fixed daily dose of 34.5 mg/kg.

Representation of uncertainty about the LD50

Acute toxicity varies between species for the same pesticide, and is usually only measured for a small number of standard test species. The LD50 for our focal species, the blue tit, is unknown, so the probabilistic assessment used a distribution to represent this uncertainty. Although the toxicity of chlorpyrifos has been tested for a large number of species this study used only the single result for bobwhite, as quoted earlier. This was done to explore the applicability of probabilistic methods to the more normal situation, especially for new pesticides, where toxicity data are available only for one or two species.

Table I. Summary of data used in case study.

Variable	Deterministic assessment	Probabilistic Model 1	Probabilistic Model 2	Probabilistic Model 3
LD50	32 mg/kg			
Probit slope of LD50	Not used in deterministic assessment	Distributions estimated from one toxicity study for bobwhite quail and historic data on many pesticides (see text)	Distributions, as Model 1	Distributions, as Model 1
C – Concentration of pesticide on food	27.8 mg/kg	27.8 mg/kg	Distribution based on field data on concentrations in insects (see text)	Distribution, as Model 2
PT – Proportion of food from treated area	1	1	1	Distribution based on radio-tracking blue tits in orchards (see text)
TFIR – Total food ingestion rate	3.3 g/day	3.3 g /day	3.3 g /day	3.3 g /day
FDR – Fresh to dry weight ratio for food	5	5	5	5
W – body weight of bird	13.3 g	13.3 g	13.3 g	13.3 g

A distribution for the LD50 was estimated in two steps using methods proposed by the ECOFRAM project (*1*). First, the mean of the distribution of log LD50s between species was estimated by adding an extrapolation factor to the LD50 for the bobwhite quail:

Mean (log LD50) = log LD50$_{bobwhite}$ + Extrapolation factor   (2)

ECOFRAM estimated this extrapolation factor by calculating the ratio between the log of the bobwhite quail LD50 and the geometric mean LD50 for 56 cholinesterase-inhibiting pesticides. The mean value of this ratio for the 56 pesticides was –0.0177, with a standard deviation of 0.38 (*1*). A normal distribution with this mean and standard deviation was therefore used in Model 1, to take account of uncertainty due to the varation of the extrapolation factor between pesticides.

Second, the standard deviation of the distribution of log LD50s between species was set to 0.428, this being the pooled standard deviation calculated by ECOFRAM (*1*) for the same set of 56 pesticides using the formula proposed by Luttik and Aldenberg (*11*). Model 1 therefore used a standard deviation of 0.428, together with means calculated using equation (2), to define a normal distribution representing uncertainty about the LD50 of chlorpyrifos for blue tits.

Representation of uncertainty about the slope of the dose-response

The slope of the probit dose-response also varies between species. Unfortunately most of the toxicity studies done with non-standard species do not report slopes, so varation in the slope cannot be analysed satisfactorily with the method used for the LD50. However ECOFRAM (*1*) presented evidence that the overall variation between species is not much greater than the varation due to other sources (varation within and between tests with the same species). Therefore, as a simple approximation, Model 1 defined a normal distribution to represent uncertainty about the slope using the reported slope (4.6) and its standard devation (1.2) from the same bobwhite LD50 study for chlorpyrifos, referred to earlier.

Estimation of percentage mortality

ECOFRAM (*1*) identified a range of options for combining exposure and toxicity, including the simple option of generating a distribution of toxicity-exposure ratios in place of the single ratio produced by the deterministic assessment. However, this study examined another approach recommended by ECOFRAM, which enables risk to be estimated in terms of the predicted mortality, because this was thought to give a more interpretable representation of the expected impact in the field. The method proposed by ECOFRAM (*1*) for this purpose involved calculating the dose required to kill a random individual bird (its tolerance), estimated as:

$$\text{Individual tolerance (mg/kg)} = LD50 \times 10^{(z \, / \, slope)} \qquad (3)$$

where
$LD50 =$	median lethal dose for species
$slope =$	slope of probit curve for LD50
$z =$	standard normal deviate.

This calculation was necessary because the tolerances of individual animals are distributed around the median lethal dose (LD50). Assuming the probit model, the distribution of the base 10 logarithms of the tolerances is Normal with mean = log LD50 and standard deviation = 1/slope (ECOFRAM report Appendix D1 (*1*)). The standard normal deviate (z in equation *3*) is a number taken at random from a normal distribution with a mean of zero and standard deviation of 1, and is used to model the distances of randomly chosen individuals from the median tolerance. The effect of this calculation is that each individual is equally likely to fall in any percentile of the distribution of tolerances.

The tolerance for each individual from equation (*3*) was compared with its daily dose, which in Model 1 was fixed at 34.5 mg/kg (as stated earlier). If the dose exceeded the tolerance, the individual was assumed to have died; if the dose was less than the tolerance, the individual was assumed to survive. This operation was repeated for 1000 individuals, each using a different value for *z*, and the percentage of individuals dying was calculated.

In order to examine the effects of uncertainty about the LD50 and slope of the dose-response relationship, the calculation for percentage mortality was repeated in 1000 simulations. Each simulation took one value for the LD50 and one value for the slope at random from the distributions defined above, used them to calculate tolerances for 1000 individuals with equation (*3*), compared the tolerances to the daily dose of 34.5 mg/kg and determined the percentage of mortalities. Altogether the 1000 simulations generated 1000 estimates of percentage mortality, forming a distribution for percentage mortality that represented the effect of uncertainty about the true values of the LD50 and slope for the blue tit.

Results

An 'exceedance curve' was plotted to show what proportion of the 1000 simulations exceeded any given level of mortality (Figure 1). For example, the lines drawn on Figure 1 show that mortality exceeded 90% in 33% of the simulations, and exceeded 56% in 50% of the simulations.

In effect, the 1000 simulations represent 1000 hypothetical species of small insectivorous birds with different LD50s and dose-response slopes. The exceedance curve can be interpreted in two ways. Most simply, it shows the proportion of small insectivorous bird species that will exceed any given level of mortality. Alternatively, if one is interested in a particular species (such as the blue tit) for which the LD50 is unknown, the exceedance curve shows the chance that its mortality exceeds any given level. For example, there is a 1 in 3 chance of mortality over 90%, and a 1 in 2 chance of mortality over 56%.

It is important to note that Model 1 did not include all possible sources of uncertainty regarding toxicity; for example it did not take account of possible differences between the dose response relationship in laboratory studies and the field. Nevertheless, the results are likely to substantially over-estimate the true level of risk because Model 1 used the same exposure estimate as the deterministic assessment, including some potentially conservative 'worst-case' assumptions. Models 2 and 3 examined the effects of replacing two of these assumptions with more realistic data.

Model 2 – incorporating variability in pesticide residues

Model 2 explored the effect of using more realistic estimates of the concentrations of pesticide on the small insects eaten by blue tits (C in equation (1)). The assumed residue of 27.8 mg/kg used in Model 1 was derived from measurements of residues on plant material, as explained earlier. This was replaced in Model 2 by a distribution based on measurements of residues on insects collected within 24 hours after spray applications of pesticides in the USA (see chapter 3.10.6.3 of the ECOFRAM report ([1])). Only data for applications to orchards (citrus and apples) were used. These data derived from 4 studies with unnamed pesticides applied at rates between 1 and 6 kg/ha. Measured residues were divided by the corresponding application rate to obtain a 'residue per unit dose' (RUD). RUDs ranged from 0.04 to 10.99 mg/kg with a mean of 2.5 mg/kg and standard deviation of 2.8 mg/kg. The RUDs fitted reasonably well to a lognormal distribution, but it was decided instead to use the

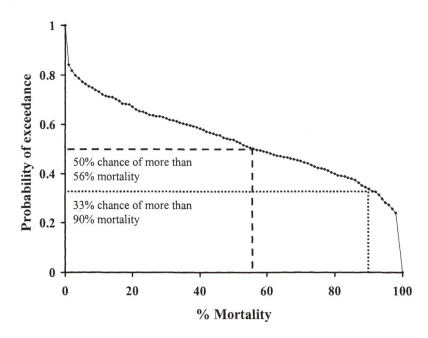

Figure 1. Probability of exceeding any given level of mortality for blue tits exposed to chloropyrifos. Results of Model 1, incorporating uncertainties regarding toxicity. (Reproduced with permission from CSL Science Review, *1999–2000. Copyright 1999 Central Science Laboratory.)*

non-parametric 'General' distribution fitted by RiskView (© Palisade Corporation), which in effect adopts the shape of the empirical distribution. Model 2 therefore sampled 1000 values at random from the distribution of RUDs and then multiplied each value by 0.96 kg/ha, the application rate for chlorpyrifos in the model scenario. The resulting concentrations (mg/kg) were then simply substituted for C in equation (1), generating 1000 estimates for the daily dose. As the RUD distribution represents variation between orchard plots, the 1000 estimates of daily dose can be regarded as representing 1000 individual blue tits foraging in different plots. Tolerances were estimated for the 1000 individuals and combined with their daily doses to produce an estimate of percentage mortality, using the same method as for Model 1. Finally the whole procedure was repeated 1000 times, with different values for the LD50 and probit slope, to obtain a distribution of percentage mortalities, again using the same method as for Model 1.

Results

The results for Model 2 show much lower mortalities than Model 1, because the range of residues is much lower. For example, the exceedance curve for Model 2 shows roughly a 20% chance of mortality exceeding 20% (Figure 2).

It is important to note that although Model 2 incorporated a distribution to represent variability in residues on insects, based on more relevant data than were used in Model 1, there are several important sources of uncertainty regarding this distribution. First, the distribution of RUDs for chlorpyrifos may not be the same as that for the unnamed pesticides to which the data refer. Second, the insects collected in the field studies are unlikely to be representative of those taken by blue tits, because they were collected by pitfall trapping at ground level whereas blue tits feed in the tree canopy. Third, the application method and environmental conditions may differ significantly between the UK and the USA. Fourth, the use of RUDs assumes a linear relationship between application rate and residue, which seems reasonable but is untested. The true residues could therefore be either higher or lower than those used in Model 2. Ideally, these sources of uncertainty should be included in the model; this could be depicted as a broad band of uncertainty around the exceedance curve for Model 2 in Figure 2.

Model 3 – incorporating variability in the foraging behaviour of birds

Model 3 explored the effect of using more realistic estimates of the proportion of their food that blue tits obtain in treated areas (PT in equation (1)).

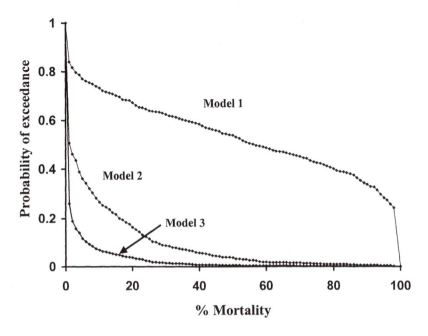

Figure 2. Probability of exceeding any given level of mortality for blue tits exposed to chloropyrifos. Results of Models 2 and 3, compared to Model 1. See text for details. (Reproduced with permission from CSL Science Review, 1999–2000. Copyright 1999 Central Science Laboratory.)

The assumed value of 1 used in Model 1 is an extreme worst-case. This was replaced in Model 3 by a distribution based on measurements in the field. Ideally, these would be measurements of the proportion of food actually taken by blue tits from sprayed orchards, but such measurements are practically impossible to obtain. Instead, it was assumed that the proportion of food taken in orchards is equal to the proportion of time birds spend in orchards, as measured by radio-tracking of 23 blue tits caught in and around apple orchards in the UK (Figure 3). These data were incorporated in the model as a non-parametric 'General' distribution fitted by RiskView (© Palisade Corporation). Model 3 therefore sampled 1000 values at random from this General distribution and simply substituted them for PT in the calculation of daily dose using equation (1). All other aspects of the model remained as in Model 2, including the General distribution for residues in insects.

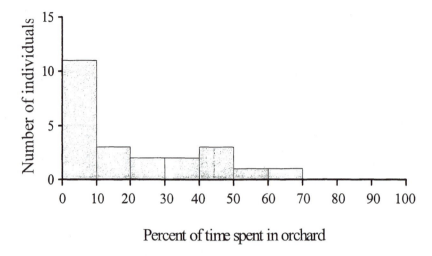

Percent of time spent in orchard

Figure 3. Distribution of time spent in orchards and therefore potentially exposed to pesticide applications by 23 blue tits, as measured by radio-tracking. (Reproduced with permission from CSL Science Review, *1999–2000. Copyright 1999 Central Science Laboratory.)*

Results

The results for Model 3 (Figure 2) show lower mortalities than Model 2, because most blue tits spent only a small proportion of their time in orchards and therefore received lower exposures. For example, the exceedance curve for Model 3 shows a 19% chance of mortality exceeding 1%. The exceedance curve is a flexible way of presenting results as it shows the probability of any given level of mortality. Alternatively, results can be tabulated for specific levels of mortality, or the average mortality over all 1000 simulations can be given (Table II).

Table II. Summary of results for Models 1-3

	Average mortality	*Probability that mortality exceeds 1%*	*Probability that mortality exceeds 10%*
Model 1	53%	0.84	0.73
Model 2	9%	0.51	0.27
Model 3	1.4%	0.19	0.05

Care is needed to interpret the results correctly. Model 3 refers to the part of the blue tit population that lives around orchards, because the radio-tracking data used in the model relate to blue tits that were caught in and around orchards. The result in the third column of Table II can therefore be interpreted as showing a 19% chance that mortality of blue tits living around orchards exceeds 1%. Note that the estimated mortalities are estimates of the overall mortality for a large number of birds chosen at random from different orchards. Individual orchards would show higher or lower mortalities, as the distribution of residues used in Models 2 and 3 mainly derives from variation between orchards (see introduction to Model 2, above). If desired, Model 3 could be restructured to produce distributions showing the variation in mortality between orchards. Alternatively, Model 3 could be expanded with additional data to include blue tits living away from orchards, in which case all the estimated mortalities would be lower.

Model 3 can also be used to make statements about other small insectivorous species of birds, provided their foraging behaviour is similar to blue tits (i.e. similar to Figure 3). In this case, the interpretation of the results in Figure 2 and Table II is slightly different: they estimate the proportions of these species that will exceed given levels of mortality. For example, the results indicate that 19% (about 1 in 5) of these species will suffer mortality greater than 1%, and the average mortality taking these species together would be 1.4% (Table II). As before, this refers to those individuals living around orchards.

It is important to note that although Model 3 incorporated a distribution to represent variability in bird behaviour (Figure 3), based on field measurements, there are several important sources of uncertainty regarding this distribution. First, it was derived from a limited sample of blue tits and orchards. Second, radio-tracking was conducted at various times throughout the season when pesticides are applied; it is conceivable that birds use orchards more intensively immediately after applications, either because spraying is a response to high insect levels, or if insects become more easily available to birds after spraying. Third, the proportion of time spent in orchards may not be representative of the proportion of food obtained there. Ideally, all significant sources of uncertainty should be included in the model; this could have been depicted as broad bands of uncertainty around each of the exceedance curves in Figure 2, and wide confidence limits on the results in Table II.

Model 3 was still very simple. It took account of uncertainty regarding the toxicity of chlorpyrifos to blue tits and variability in residues and bird behaviour but, as noted above, it did not include all the uncertainties affecting these parameters. The parameters TFIR, FDR and W in equation (1) were treated as fixed; in reality they are affected by both variability and uncertainty. Furthermore, the model omitted other potentially important processes including the potential for birds to actively avoid contaminated food (which could reduce exposure) and the potential for exposure by non-dietary routes (which could increase exposure). Finally, it was assumed that all the parameters were independent, which may not be true. All these things could, and ideally should, be included in the model unless there was evidence that they were unimportant.

Discussion

As would be expected, the level of risk decreases markedly from Model 1 to Model 3, because worst-case assumptions regarding C and PT are being replaced by distributions based on more realistic data (Figure 2 and Table II). A similar decrease could be shown by using point estimates of C and PT (e.g. means or specified percentiles) in a refined deterministic assessment, but this would not reflect the full range of variability in these parameters. The ability to incorporate variability is one of the key advantages of the probabilistic approach, and provides a more complete description of risk for the decision-maker.

Probabilistic approaches also provide a means of incorporating uncertainty concerning the model parameters, as illustrated by the treatment of toxicity in the models. Similar methods could be applied to other forms of uncertainty, to provide the decision-maker with a clearer understanding of how they affect the assessment outcome; this could be represented either by putting confidence bands around exceedance curves or confidence limits on numerical estimates of risk. This is another substantial advantage over deterministic methods, which account for uncertainty using simple assessment factors that generally have not been derived from an objective analysis of uncertainty. However, more work is needed to apply uncertainty analysis more comprehensively in probabilistic assessments for pesticides, as is illustrated by the many uncertainties not quantified in this paper.

Another important advantage of quantifying uncertainty, not explored in this paper, is that it reveals which sources of uncertainty have most impact on the risk estimate. This provides an objective basis for specifying which additional data are needed to refine the assessment, and should increase the cost-effectiveness of the assessment process.

It is frequently suggested that probabilistic assessment requires more data than current approaches. In fact, as the example in this paper shows, a limited probabilistic assessment can be conducted with the same minimum dataset that is currently used for the initial deterministic assessment. Many more uncertainties need to be incorporated, as already mentioned, but some methods of uncertainty analysis are specifically designed for working with limited and/or subjective information, without requiring every parameter to be quantified precisely (4). More work is needed to explore the applicability of these methods to pesticide risk assessment.

Conclusions

Probabilistic risk assessment methods offer substantial advantages for assessing the impacts of pesticides on birds, and may provide a practical solution to many of the difficulties which are encountered using current procedures.

285

However, the present example was exploratory and was not intended as a model for regulatory purposes. Substantial further work is required to implement probabilistic methods and achieve a consensus on how they should be used. Amongst other priorities (4), there is a need to be more comprehensive in incorporating uncertainty, and to explore the applicability of methods designed for use with limited and/or subjective information.

Acknowledgements

The author is grateful to ECOFRAM project members, Mark Clook, Dwayne Moore, Scott Ferson and many other individuals for ideas and discussion; to Monty Mayes for encouragement; to Pierre Mineau for providing toxicity data; to Alain Baril for help with the analysis of uncertainties in Model 1; to Dave Fischer for providing residue data; to colleagues at CSL for the data in Figure 3; and to the UK Department for Environment, Food and Rural Affairs for funding.

References

1. *ECOFRAM Terrestrial Draft Report*; US Environmental Protection Agency Ecological Committee on FIFRA Risk Assessment Methods (ECOFRAM); Available at www.epa.gov/ecotox/, **1999**.
2. *ECOFRAM Aquatic Draft Report*; US Environmental Protection Agency Ecological Committee on FIFRA Risk Assessment Methods (ECOFRAM); Available at www.epa.gov/ecotox/, **1999**.
3. *A Progress Report for Advancing Ecological Assessment Methods in OPP: A Consultation with the FIFRA Scientific Advisory Panel. Overview Document*; US Environmental Protection Agency; Available via www.epa.gov/scipoly/sap/ (see page for meeting of April 2000), **2000**.
4. *Probabilistic risk assessment for pesticides in Europe: implementation and research needs. Report of the European workshop on Probabilistic Risk Assessment for the Environmental Impacts of Plant Protection Products (EUPRA)*; Hart, A. (ed.). Central Science Laboratory, Sand Hutton, UK. Available at www.eupra.com, **2001**, 109pp.
5. *CRC Handbook of avian body masses*. Dunning, J.B. CRC Press, Boca Raton, Florida, USA, **1993**.
6. Nagy, K.A. Field metabolic rate and food requirement scaling in mammals and birds. *Ecol. Monographs* **1987**, *57*, 111-128.
7. Kenaga, E.E. Factors to be considered in the evaluation of the toxicity of pesticides to birds in their environment. *Environmental Quality and Safety* **1973**, *2*, 166-181.

8. EPPO/CoE. Decision-making scheme for the environmental risk assessment of plant protection products: terrestrial vertebrates. *EPPO bulletin,* **1994**, *24*, 37-87.
9. Hill E.F.; Camardese, M.B. Toxicity of anticholinesterase insecticides to birds: technical grade versus granular formulations. *Ecotoxicology and Environmental Safety* **1984**, *8*, 551-563.
10. EU Directive 91/414/EEC, Annex VI.
11. Luttik, R.; Aldenberg, T. Extrapolation factors for small samples of pesticide toxicity data: special focus on LD50 values for birds and mammals. *Environmental Toxicology and Chemistry,* **1997**, *16*, 1785-1788.

Indexes

Author Index

Subject Index